Paleolithic Diet

Digging Deeper In To The Original Human
Diet and Paleo Recipes

Lindsay Sullivan and Bill Summers

Table of Contents

Introduction

Known among its advocates as the Caveman's Diet or the Stone Age Diet, the Paleo Diet is more than just an ordinary eating regimen. It is in essence a lifestyle that mimics the eating habits of our ancient ancestors who lived during the Paleolithic era or the stone-age era. The dietary regimen includes limiting your food intake to the type of food groups our prehistoric ancestors ate using similar but contemporary foodstuff. It is the latest and hottest trending topic to hit the health and fitness scene of late.

Dismissed as a passing fancy and a fad by some so-called nutritional experts, the Paleo Diet has nevertheless continued to rise in popularity gaining a growing number of advocates who have found wisdom in going back to the basics of primal living.

Whether it will just be a temporary trending diet fad that may die down soon enough or it will become a mainstream dietary plan, only time will tell. In the meantime, the ever increasing number of Paleo advocates who'd swear to its efficacy in gaining great health and perfect physique is reason enough for us to give the concept more than just a cursory look.

This 3-part eBook will give you an insight to the wonderful world of the Paleo Lifestyle and guide you on how you can adapt to this primal style of eating and living to your own. It will debunk all the myths and lies being circulated around to discredit its incomparable nutritional value. It will teach you how to get healthy with a primal diet rich in protein and animal fat but low in carbohydrates the way our primal ancestors did.

For those who are new to the Paleo concept, this eBook will outline the logic and the compelling reasons why we should return to primal eating. Whatever it is you are searching for - an effective way to lose weight, practical approaches to physical conditioning, or a cure to some illnesses, this book has a wealth of useful information to help you achieve your objectives.

Part 1 Understanding the Paleo Diet

Chapter 1: What is the Paleo Diet?

The Paleo Diet is basically a dietary concept built on the belief that by eating in the like manner our stone-age ancestors did, and limiting our food intake to the kind of food available to them then, we will become leaner, meaner, and healthier. It is more than just a bunch of well concocted recipes. It is a whole lifestyle which also involves eating similar food in the most natural state possible.

The Paleolithic Era and the Human Genome

The Paleolithic era is what is commonly referred to as the stone-age era. It marked that period in the prehistoric human history where man started learning how to craft various tools out of stone. The word 'Paleolithic' is derived from two Greek words which mean "Old age of the stone" or Stone Age for short. It was also the era when men discovered how to make fire and started cooking the food they ate. It was an era that began about 2.6 million years BP (Before Present) and ended just 10,000 years ago period when agriculture and animal husbandry became known to man.

It is largely believed that people during the stone-age era were basically hunter-gatherers who survived by banding together in small groups to hunt wild animals and gather edible wild plants for subsistence. These prehistoric men survived on  minimally processed natural food for millions of years and the human body was thought to have adapted to it perfectly. The human genome is believed to have evolved already programmed to getting its nutrients from natural sources after consuming the same minimally processed food for millions of years.

The advent of agriculture and animal husbandry just 10,000 years ago has brought profound changes to the way people ate and the kind of food they had. As man's knowledge grew, they also learned different ways to produce and process food more efficiently. Food started to be processed in ever increasing scale quickly replacing natural food sources with which man has been used to for millions of years. The industrial revolution that ensued in the same era further hastened man's preferential shift to processed food as they became readily available and easier to acquire. Mass production of processed food became widespread and their consumption was widespread and unprecedented. The people's choices of food have by now dramatically changed. They started to prefer processed foodstuff over everything natural and the slightly processed food stuff which the human body was used to eating for millions of years became a thing of the past.

It took some time before a group of nutrition and health experts started to recognize that the many profound changes which happened in and around our environment and ended the stone-age era have also created discordance between our naturally evolved ancient human genome and the contemporary processed food of modern times. They believe that the human body has been forced to adapt to the changes abruptly and unnecessarily - thus it has become naturally strained.

They believed that this discordance may have caused many of the illnesses of modern times as the body have not fully adapted to contemporary food stuff. 10,000 years, to them was not enough for the human body to adapt readily to processed food and supplements which dominates the modern urban diet. In the first place, it took 2.6 million years for the human genome to evolve towards the Paleolithic diet and was practically programmed to it.

So, what did the Paleolithic people really eat?

 What our prehistoric ancestors actually ate is something modern man won't really be able to specifically determine. None of us lived during those times and therefore, it would be hard to pinpoint exactly what they actually ate then. We cannot possibly go back in time, can we?

At the very least, we can merely make an educated but well-founded guess based on whatever information we have uncovered from the past. Based on numerous research studies, the type of food prehistoric people ate were largely limited to whatever was available in their geographic locations during any given time. What they are can be deduced through meticulous scientific studies and advanced laboratory analysis of the prehistoric bones and dentures of Paleolithic people. Results from such studies served as basis for our assumptions of what constitutes the primal diet.

We can also make plausible assumptions on what constitutes the primal diet through sheer logical reasoning. By studying the type of food we have today which couldn't possibly have existed during the Paleolithic era, we should more or less be able to determine what type of food prehistoric people never ever consumed through the process of elimination. By striking out contemporary foodstuff which could not have possibly existed during the stone-age era from our food list, we should be able to come out with an idea on what our ancestors ate.

For example,

- They had no dairy products then as animals were not yet domesticated. It would be utterly impossible and even too risky to milk wild cows (if they already existed) or other lactating wild animals.
- Agriculture wasn't existent then and therefore Paleolithic people hardly had cereal grains. Whatever grains they may have had may have been gathered from plants that grew wild in the fields which should be in largely limited quantities.

- They never salted their food since they didn't have salt too at that time. There is no documentary evidence existing today that shows stone-age people mined salt during their era. The only possible thing they could have done then was to dip their food in salt water.
- Sugar was not yet available too at that time and the only possible sweetener they may have used is wild honey which we can assume was also hard to find at that time.
- Lean meat from wild animals was their common fare which means their diet had higher protein content compared to today's diets.
- Their consumption of carbohydrates is also low but rich in fiber compared to modern diets as their carbohydrate source come from wild fruits and plants existing at that time most of which are non-starchy and therefore have lower carbohydrate content.
- They didn't have trans-fats like what we usually get from processed foodstuff today. What they had were omega3 fats, polyunsaturated fats and healthy monounsaturated fats from lean meat, fish and seafood.

Chapter 2: What constitutes the Paleolithic Diet then?

Presumably, the original prehistoric human diet consisted of untamed plants and various wild creatures that they were able to gather or hunt down. We can safely assume that the prehistoric diet consisted of lean meat from free-ranging, grass fed wild animals, non – starchy wild fruits, nuts, and vegetables which are low in carbohydrates, fish, and seafood. There were no dairy products, refined sugar, cereal grains, legumes nor processed foodstuff back then, and so we should eliminate them from our primal diet food list.

The significance of the Caveman's Diet to Man's Health

Researchers who have long been studying this so-called 'caveman's diet have noted a most significant thing which inspired them to pursue the Paleo Diet deeper. Cavemen were never afflicted by the chronic illnesses and diseases pervasive among men today. Significantly absent among prehistoric men are such health conditions like cardio vascular diseases, cancer, diabetes, gout, osteoporosis, varicose veins, macular degeneration, autoimmune diseases, glaucoma, and a lot more.

Known researchers like Dr. Lorain Cordain who is a recognized authority on Paleo Diet have linked this amazing phenomenon to the caveman's diet. He has studied it for decades now together with his associates and they have firmly concluded that prehistoric men never suffered from the same illnesses and diseases contemporary men suffer from because of the different kind of food they eat. And, the food they eat is basically low in carbohydrate, rich in protein, and full of phytochemical nutrients from natural sources.

 The Paleolithic diet in contemporary terms is made up mainly of grass-fed pasture elevated meat, fish, seafood, fruit, organic vegetables, root crops, and nuts. It does not include grains, beans, dairy products, refined sugar, salt, and processed oils.

Chapter 3: The Contemporary Western Diet

Its Origin, Evolution, and its Implication to our Health

 As prehistoric men began to learn how to domesticate wild plants and animals, agriculture and animal husbandry became man's most important productive food endeavors. With it came profound changes on how and what people ate. It significantly transformed people's lifestyle particularly their eating habits. And when the ensuing Industrial Revolution introduced novel ways of producing and preparing food  the cave man's diet soon became a thing of the past.

Significantly, the contemporary Western Diet has considerably altered what used to be the normal nutrient intake of the human body which for millions of years had been accustomed to, beginning from the stone-age era. While the usual protein content of today's contemporary diets is only 15%, the human body is used to having an average of 19% to 35% protein from the cave man's diet.

Contemporary diets have also been totally dominated by refined grain and refined sugar which is today's main source of carbohydrates for energy and a staple in modern urban diets. Unfortunately, studies have shown that both refined sugar and cereals have high glycemic indices and can raise the blood sugar levels a lot. This is in stark contrast to the low carb, non-starchy fruits, nuts, and vegetables consumed by our stone-age ancestors.

Based on most recent studies, carbohydrates in modern urban diets supply 39% of the body's energy needs. Both have high glycemic loads which induce spikes in blood sugar levels. The high incidence of chronic diseases affecting a large number of Americans today has been attributed to this type of high carbohydrate diet.

More concerning are the following health figures. According to the most recent statistics, 65% of adults in the U.S. aged 20 and above are obese or at the very least overweight. 64 million Americans suffer from cardio vascular disease while another 50 million have hypertension. 11 million Americans suffer from type 2 diabetes and this is directly linked to high concentrations of sugar in the blood plasma due to prolonged intake of foodstuff with high glycemic index. 25% of deaths in the country are due to cancer and 1/3 of these deaths have been linked to nutritional factors. These statistics are reasons enough for us to take a closer look at the kind of food we eat.

 The fact that men started consuming the modern urban diet with high glycemic loads and low in protein barely 200 years ago makes it suspect. For millions of years before then, it was just the Paleo Diet.

Prior to the Industrial Revolution, grains and sugar were roughly milled using only stone grinders. Today, mechanical grinders are used to finely mill them, losing much of their natural nutrients in the process. With the Western diet practically dominated by refined cereals and sugar, we are practically killing our bodies little by little. The pandemic rise of cardio vascular diseases in the last 200 years is particularly disturbing and has been linked to the Western diet.

There is therefore every reason to believe that man's shift to the modern urban diet had everything to do with the emergence of chronic diseases that afflict modern man today. These illnesses never affected men until the most recent times when progress in agriculture came by leaps and bounds, and inorganic fertilizers were developed to produce bountiful harvest; until animal husbandry introduced new methods of herding livestock in contained environments, feeding them corn and cereals instead of allowing them to graze freely and feed on grass; and until the introduction of modern methods of processing food and supplements which use mostly synthetic ingredients produced in laboratories.

Interestingly, as more and more people are held sway by the easy to acquire and easy to prepare contemporary urban diet, the incidence of chronic diseases and illnesses dramatically rose in tandem. This led Paleo advocates to conclude that the modern urban diet actually gave birth to the so-called 'diseases of affluence'.

They theorized that much of what ails men today can be linked to their diet compounded further by a sedentary lifestyle which progress forced into the modern man. Men today drive to get to work instead of walk. They use the elevators instead of the stairs. They have largely cut back on physical activities and this exacerbated the detrimental effects of the modern urban diet to the human body.

The Paleo advocates argue, with good reason of course, that the human body which for millions of years had been so used to being sustained by nutrients gathered from natural food sources encounters difficulty in adapting to the contemporary diets and therefore had become unnecessarily strained. As a result, the natural immune systems of some people started developing negative reactions in the form of illnesses and diseases which we now commonly refer to as the diseases of affluence.

Chapter 4: The Transformative Logic of the Paleo Diet

The underlying logic upon which the Paleo Diet is based is built on the assumption that the genetic composition of modern man has been programmed towards the diet of our stone-age forefathers and it has not changed since. For more than 2.5 million years, man has had the same natural diet consisting of wild plants and animals so much so that the human body has already become accustomed to it. The human genome is believed to be already programmed towards this type of diet and the agricultural and industrial revolution in the last 10,000 years has not transformed it. This is the transformative logic on which the Paleo Diet is based.

The premise states that the genetics of modern man had scarcely changed through these years even after the advent of modern agriculture and animal husbandry and despite the large scale modernization of food processing. They believe the 10,000 years covering the time when man first learned how to cultivate plants and domesticate animals up to the present when the modern urban diet evolved wasn't enough to reprogram the human genome. Besides, man started consuming the modern urban diet consisting largely of processed food barely 200 years ago such that it would be impossible for the body to totally reprogram what took it 2.5 million years to program. Worst, the human body is reacting negatively to the modern urban diet as manifested through various chronic diseases that now plague modern man.

It was a Gastroenterologist by the name of Walter L. Voegtlin who first toyed with the concept of the stone-age diet sometime in the mid-70's. Since then, a number of books and academic journals have been written about it. Together, they show that there is increasing evidence to prove that a diet consisting of lean meat, fish, fruit, and vegetables similar to the diet of our prehistoric ancestors can prevent the so-called diseases of affluence modern men are now commonly afflicted with.

The Paleo diet grew in popularity and as its acceptance was just getting a kickstart, a few authors and scientists tried to introduce some modifications to this concept confusing some advocates in the process. Others questioned the validity of its so called transformative logic without offering proof to negate the concept. Never-the-less, the Paleo diet still captured the imagination of a growing number of health and fitness advocates and it now appears it is here to stay.

Followers of the Paleo Diet believe that to ensure health and well being, and remain free from diseases of affluence, modern man has to adapt to a diet that resembles the diet of his pre-historic ancestors.

This transformative logic of the Paleo concept did not sit well with some quarters. A number of dieticians and anthropologists, especially those who champion other modern day dietary programs have questioned its validity claiming that much of the Paleo concept is merely based on guess work.

True enough, all of the health professionals who championed the Paleo Diet advocacy like Walter L. Voegtlin and Dr. Lorain Cordain did not live yet during the Paleolithic era to be able to have a firsthand account of what the prehistoric people had for their meals. Essentially, we can say that most of the contentions they wrote about were mere suppositions. However, no matter how presumptive the Paleo concept may be, it is not totally baseless.

The truth is much of the Paleo assumptions were culled from well documented studies and extensive laboratory analyses which include in depth evaluation of the dental function morphology of Homonin Fossil Records and Paleo-environmental modeling among others. The principles are scientifically sound, the logic highly reasonable.

We are not however, going to delve into the more scholarly aspects of the primal diet. What we are more interested in is whether or not following the said diet will help us remain healthy and free from the dreaded diseases and illnesses which have been linked to the modern urban diet.

For us, we can offer no better proof than the various contemporary hunter-gatherer tribes still existing today in various parts of the globe who continue to subsist on a similar primal diet. There is the Hazda Tribe of Tanzania, the Veddas of Sri Lanka, and the Mbuti pygmies in Congo, the Guyaka Indians of Paraguay, the Eskmos of Alaska, the Aborigines of Australia, and the East Coast American Indians. The chronic diseases of affluence modern man is prone to are practically absent among these tribes. They are the living proof of how primal diet is important in keeping the human body healthy and disease free.

If you want more scientific basis showing the efficacy of the Paleo Diet, here's one incontrovertible proof for you which Paleo detractors try to simply ignore because they cannot disprove it.

Researchers from the University of California San Francisco (UCSF) headed by Dr. Timothy White and Dr. Linda Frasseto did an actual test on a group of people who were all considered as unhealthy or have an ailment of some sort. The group was given a Paleo Diet consisting of lean meat, fresh fruits, fish, nuts, and vegetables.

After just two weeks on this diet, the blood pressure of everyone in the group went down dramatically, while cholesterol and triglyceride levels dropped an average of 30 points – a feat which according to Dr. Frasseto would have taken drugs like Statin and other cholesterol lowering medications six months to achieve.

There are countless other stories of how the Paleo diet helped ailing individuals regain their health many of which were never published. Countless accounts of how the diet helped improve the physique and the performance of professional athletes also abound. The diet is not a mere fad as most people are made to believe by its detractors. It is a lifestyle. In fact it is well on its way to becoming the lifestyle of the future – the very reason why proponents of other dietary regiments try to discredit its real value by tagging it as an uncivilized way of eating and living. Unfortunately, they have yet to disprove its efficacy which many of its advocates would swear to even with their own lives.

Chapter 5: The Paleo Diet versus the 21st Century Western Diet

What our prehistoric ancestors had were natural, unprocessed food which they forage or hunt down within their immediate environs. The modern urban diet on the other hand is mostly processed foodstuff dominated by refined sugar and cereals and synthetically prepared ingredients. The comparison is clearly a toss-up between what is natural and what is synthetic. As to which one is better, we shall leave this up to your better judgment.

But what has been construed as the Paleo diet actually has two to three times more fiber than the typical average Western diet. It has two times more polyunsaturated fats and monounsaturated fats and four times more Omega 3 fats. The more significant thing with this primal diet is that it has 30% to 40% less saturated fats than the modern urban diet. Its protein content is two to three times more than the average contemporary urban diet.

Since the Paleo diet contains moderate amounts of beneficial fats and carbohydrates with low glycemic indexes plus a load of useful phytochemicals, it is the perfect diet to prevent weight gain and avoid cardio vascular diseases. On the other hand, there are voluminous proofs which link the modern urban diet to high incidence of pandemic cardio vascular diseases and obesity today.

The monounsaturated fat content of the Paleo Diet comes mostly from nuts and has been proven to protect the heart from cardio vascular diseases in at least six clinical studies. The omega 3 fat content is also higher in the primal diet than in the meat from today's domesticated animals since they are grain or corn fed and not fed with omega-3 rich grass. Omega 3 fat also has cardio protective properties. The carbohydrate content of the primal diet has lower glycemic index than the carbs from contemporary diets since they basically come from non-starchy fruits and vegetables.

It is undeniable that the human body remains genetically adapted to the primal diet. While it is impossible to replicate the diet exactly as it existed during the prehistoric era, it can serve as the guideline to design effective diet interventions to protect man from the incidence of cardio diseases.

Chapter 6: The Pros and Cons of the Paleo Diet

Since the Paleo diet gained prominence, it attracted advocates as well as detractors. Naturally, the advocates will only sing praises and heap accolades for this new dietary regimen. On the other hand, its detractors will be quick to point out perceived flaws in its logic. For your consideration, we are enumerating here all the arguments pro and against the Paleo concept coming from both sides. We shall leave the final call to your better judgment.

The Pros

Paleo diet advocates proclaim that the diet provides a host of significant health benefits among which are:

1. The Paleo diet protects against weight gain. Since the diet prescribes only non-starchy fruits and vegetables as its main source of carbohydrates and not from grains and sugar which have high glycemic indexes, the insulin levels in the blood are lowered thus preventing carbohydrates from being converted and stored as fat. It actually leads to significant weight loss in the long term.

2. Despite having high levels of saturated fats, the Paleo diet has been shown to improve blood lipid profiles. It increases the good cholesterol (HDL) levels while decreasing the TG or triglycerides thus, effectively protecting the heart from atherosclerosis and stroke. It also has been shown to convert low density bad cholesterol (LPL) into high density good cholesterol (HDL).

3. The Paleo diet is gluten free since it leaves out wheat and other cereals from whence gluten is made from. It enhances digestion since the food intake is limited to the type of food the human body has been accustomed to for millions of years.

4. The Paleo diet eliminates blood sugar spikes and maintains stable energy levels in the body. You generally won't experience afternoon fatigues as what happens often when you eat a lot of cereals and sugar during the day.

5. It prevents bloating and promotes well being since there is more fiber intake in the salt less diet. All of the food and beverages in the Paleo diet are organic. People on the Paleo diet feel better, sleep better, and are not prone to depression.

The Cons

Critics of the Paleo diet have not wasted time to post their criticisms. Among the many objections they have so far published are the following:

1. The Paleo diet is too difficult for the simple person to follow as it entails tremendous change in one's lifestyle. It would take herculean efforts to see it through to success. With so many food restrictions, it will require changes not only in your eating habits but in shopping for food items since you need to select only organic food products and meat from livestock that has been raised and grass fed in pasture lands. It will entail taking a closer look at food labels to make sure they contain only natural and organic ingredients.

2. The shift to a Paleo lifestyle may be more expensive than usual. The fruits and vegetables have to be organically grown and this definitely will cost you more than the regular fruits and vegetables sold anywhere. The meat must be from livestock that has been grass fed or fed with corn or grains. Paleo detractors also claim that the Paleo food list includes foodstuff that are not only in short supply but are also more expensive.

3. Critics of the Paleo diet feels that by leaving out grains and cereals, the diet is in effect depriving the body of much needed fiber intake and carbohydrates. They consider the Paleo as an unbalanced diet. They also feel it is absurd to use contemporary foodstuff to reconstitute man's original ancient diet. They believe that Paleo advocates would have difficulty to stay on the diet since critics believe that because it is hard to locate purely organic and natural food sources. They predict that Paleo advocates are going to be discouraged and are likely going to abandon the diet not long after they have tried it.

4. There is however, a common ground among the advocates and the critics of the diet. Both parties do not question the fact that man should eat foods as natural and as fresh as possible. This is actually the very essence of the Paleo lifestyle and it appears that the above criticisms are more like excuses not to adapt the diet than reasons to invalidate the efficacy of the diet.

Chapter 7: Correcting the Misconceptions about the Paleo Diet

Contrary to what the critics say, the Paleo diet is the easiest and the simplest diet to follow compared to other dietary regimens. It requires no calorie counting. It just needs a great resolve to stick to regular organic food intake and to avoid processed foodstuff and synthetic ingredients. It is the healthiest too.

Dieters are simply fond of making excuses when they can't resist the urge of backsliding to their old eating habits and succumb to the urges of their old cravings for things sweet. Let us try to analyze the many misconceptions about the Paleo diet to better appreciate its benefits.

Is it a Low carb, fad diet?

Critics say the Paleo Diet is just another fad diet that is low in carbohydrates and therefore unbalanced. Nothing can be further from the truth than this baseless allegation. The Paleo diet is not low in carbohydrates. Rather, its carbohydrate content has low glycemic indices since they come from non-starchy fruits and vegetables. There is a big deal of difference between having low carbohydrate content and having carbohydrates that have low glycemic indexes.

The Paleo diet encourages you to consume a lot of fiber rich fruits and vegetables which are its main sources of carbohydrates for energy. The modern urban diet on the other hand relies a lot on sugar and cereals for energy which unfortunately causes spikes in blood sugar levels.

It can't be fad diet either because man has been consuming this kind of diet even long before the advent of agriculture. It is in fact man's survival diet since it has been able to subsist on this diet disease-free for millions of years.

Is it difficult to follow?

The most significant thing about Paleo is the fact that it entails no guess work and requires no calorie counting and constant monitoring. It is a no fuss dietary plan that does not require you to constantly make mathematical calculations or eat in the zone to make sure you are not overstepping through programmed dietary food intake.

What about Paleo food sources? Are they hard to find?

Again, contrary to the allegations of Paleo critics, Paleo food is available in a number of food outlets nationwide. Two of the more prominent food outlets where you can buy organic food are Whole Foods which has over 340 outlets nationwide and Trader Joe's which has 395 outlets in 9 states. There are also a growing number of Paleo friendly restaurants and food outlets all over the country. Natural, organically grown fruits and vegetables, and meat from grass fed livestock are not really as difficult to find as Paleo critics would like you to believe. There are even several reliable online food shops selling Paleo food items to make shopping easy and convenient.

Part 2 - Adapting to the Paleo Lifestyle

Chapter 1: Eat Like a Caveman

 Adapting the Paleo Lifestyle to your own involves eating the way the cavemen of prehistoric times did. No, it is not about getting inside a cave and dining there. It is not about scavenging or hunting for your own food since you can use contemporary food as substitute.

It is about eating contemporary food that will not cause insulin spikes. A spike in insulin in the blood actually tricks our body to store energy as fat resulting in weight gain.

It is quite unfortunate that the modern man's diet is loaded with sugar and refined cereals. Many people today are in fact so used to taking cereals for breakfast, packed sandwiches for lunch, and different pasta preparations for dinner. They even often supplement this with snacks in between meals consisting of fries, chips, and soda.

They couldn't care less whether or not the continued consumption of processed food rich in refined sugar and cereals will make them gain weight fast, and over time it may even lead to diabetes or in the worst case scenario to cancer.

Another thing that led to modern man's current predicament is his mistaken belief on what constitute a healthy diet. People have been made to believe that eating low-fat foods high in carbohydrates will help keep the heart healthy as well as prevent weight gain.

This misconception is actually the culprit. It is not eating less fat and lowering carb intake which will prevent insulin spikes. You'll only be starving the body of much needed nutrients that way. The primal diet corrects this misconception with a diet that contains more healthy fats and non starchy fruits and vegetables instead. It is not about limiting fat intake. It is about avoiding the use of trans fat and harmful omega-6 fats from soy oil or corn oil which modern man have been using extensively in preparing his food every single day.

This means eating like a caveman will leave you no choice but to forego fast food and convenience style food items. The caveman's diet does not have room for refined sugar and cereals as well as industrial oils. It is about eating food in its most natural form.

Going Primal Requires Resolve

One must have the right mindset before adopting the Paleo lifestyle. It is essential that one must be resolute to avoid backsliding into your old rotten ways. Primal living is more than a diet or a workout program – it's a total lifestyle change.

You need to realize the importance of eating and moving in ways our body have been optimally designed for. And it requires one to be unwavering when he starts going primal. It will mean abandoning many things you've grown comfortable with. It would be like turning yourself off from what used to be your comfort zone and recharging yourself with a new Paleo Lifestyle.

You also need to be more physically active

One of the ill effects of progress is it had made man lazy and less physically active so much unlike our prehistoric ancestors who are constantly on the move hunting for food or looking for ways to protect themselves from the weather. Modern man spends the most part of his waking time sitting in an office or watching television at home. Everyday he'd ride his automobile to work. Weekends is holy to him as it is the only time in a week he usually wake up late and not worry about being late for work.

Modern man has chosen to adopt a lifestyle which actually slowly erodes his muscular health and leads to cardio vascular degeneration. He needs to break from the drudgery of his day to day routine by going out under the sun more often and doing outdoor exercises.

Chapter 2: What You Need to Give Up To Be Paleo

So far, you know that you have to give up refined sugar and cereals as well as legumes to be Paleo. But there are many more things you have to give up when you embrace the Paleo Lifestyle. Sacrifices will have to be made including venturing out of what used to be your comfort zone and foregoing many things you love and have been accustomed to having. Among the things you have to give up are the following:

Avoid the fast food chains

Fast food and most other convenience food items are high in trans fat and Omega-6 fats which cause inflammation. If hamburger and fries have become part of your routine, it is time to shed the habit.

Stop making holidays as food events

Christmas, Thanksgiving, Easter, Birthdays, Anniversaries, etc. have always been a food event. We have been accustomed to celebrating such events by preparing sumptuous meals. It will be difficult to break tradition but this needs to be done for you to be Paleo. Inside of a food fest, you can suggest a trek to the park or something.

2. Walk more, Ride less

Your feet are made for walking. Try to walk as often as you can and whenever time permits. Leave your car behind whenever you can and walk instead. Don't take the elevators and use the stairs. It is good for your heart. Whatever ways you can incorporate walking or any physical exercises to your leisurely modern lifestyle will be a welcome change and is definitely going to boost your health.

3. Resist the urge

It is really difficult to eat differently from the rest of your friends. You may feel alienated or odd and the temptation to give in and eat non Paleo food with them is simply compelling. However, there are no if's and but's with the Paleo lifestyle and you need to resist all the urges and instances of once more splurging on unhealthy diet. Think about your own health in the coming years. Be strong in your resolve and you may even gain your friends' respect for your new lifestyle.

4. Prepare yourself for the long haul

Never think that Paleo is a miracle solution to uncontrollable weight gain. If you expect to lose pounds overnight you may end up disappointed. It is a lifestyle which you need to embrace for the rest of your life. The results may not be dramatic but they will surely come. They may not be immediate but positive results will come and the changes will be permanent.

5. Don't take Drugs to cure your symptoms

Throw all the medications in the house out of the window. If your doctor always prescribes drugs for any symptom, it is best to consider looking for another one. Remember that what ails man is always linked to the kind of food he takes and this is something your doctor must always consider. If you take up the Paleo challenge, you will actually help yourself get rid of all these symptoms. You just need to be firm and resolute in adopting this new lifestyle.

6. Stop being a creature of habit

If you used to bring out the kids to dine out in some fancy fast food chain every weekend or cook for them pancakes when they come home from school, it is time to break the routine and prepare a more natural, healthier meal this time around.

7. Offer no excuses

One of the common excuses people use to justify eating convenience food is a hectic working schedule. Drop the excuses. It is time to make firm changes on your food choices if you wish to pursue your health and fitness goals seriously.

8. Go to the Farmers Market Instead

You may not be able to avoid going to the supermarket for your house needs but be sure to avoid the middle section where most of the junk foods are. Better still, you should visit the farmers' market near you more often than the supermarket. They'll have plenty of fresh fruits and vegetables for you there.

9. Get more Shine from the Sun

Don't avoid going under the sun. It is healthy for you. Go outdoors as often as you can. There is no better source for Vitamin D than from sunshine.

10. Lessen your dependence on modern gadgets

Instead of losing sleep surfing the net or playing online games with your iPad in bed, tuck it away and have a good night's rest. Instead of jogging or running around on specially designed shoes that can absorb shock, run barefoot. Weird as it may seem but it is really healthy for your body.

Chapter 3: Blending your 21st CENTURY Lifestyle to the Paleo Diet

One of the most difficult things in incorporating changes to your 21st Century lifestyle is in adapting the said changes to replace established habits which don't conform well to the new lifestyle. You need to recognize which ones can be detrimental to your efforts to achieve good health and well being so you can do something to effect the changes. Here are some of them:

- If you're a coffee lover and taking your morning coffee has become a daily ritual, you can go ahead with it but you must cut back on sugar and cream. Better still take it purely black coffee – no cream and sugar at all.
- If your work keeps you busy all the time, you should survey the area around your work place and identify the food outlets which serves Paleo like foodstuff. Never leave this to chance because if you do your plan will fail since you are likely to patronize non Paleo food outlets when hunger pangs get the better of you before you can find a restaurant that is Paleo friendly.

- If you bog down once to your cravings for the usual food then you are likely to go back to your old eating habits and destroy your Paleo plan altogether.
- If you are a cake and pastry enthusiast, you have to let go of your sweet tooth. Refined sugar and dairy products have no place in the Paleo diet. You need to work hard to eliminate refined sugar and flour from your lifestyle.
- Remember that there is no room for compromise once you embrace the Paleo diet. Neither should there be space for flexibility as it will destroy the diet completely since you will start losing control over your eating habits.
- If you are an avid social creature, you must also cut back on your night life or abandon it altogether. A very active social life opens you to more temptations which will likely force you to break your Paleo diet.
- Don't eat food that uses canola oil, corn oil, soy oil or other oils made from grains and seeds. Restaurant food usually makes use of these oils and therefore should also be avoided.

- Butter, lard and tallow, unrefined coconut oil, and olive oil should be used instead. Do not be misled  by claims that saturated fats are bad for your health. They are in fact good to your heart and to your over-all health. We shall be discussing this in a separate section.
- Keep a tab on what you eat every day. Start your own food diary and record not only the food you eat but also how you felt after each food intake. The data will help you continue to adapt to the Paleo lifestyle more easily.

Our prehistoric ancestors never had the same flood of food temptations you will be faced with when you try to change over to the Paleo lifestyle. They had no choice because their very existence depended on it. Besides, it was the only thing they had and so it wasn't really hard for them. The 21st Century man on the other hand has a lot of tempting food choices to select from each single day. He must therefore be prepared to make concessions and should be willing to give up some of his favorite food items if he is truly serious in adopting the healthy Paleo lifestyle.

Chapter 4: Your Paleo Shopping List

It would be too simplistic to say that the Paleo diet consists of lean meat from grass-fed, free ranging livestock and organically grown fruits and vegetables. However, this could be vague for people who will be trying the Paleo for the first time so we prepared a Paleo shopping list to guide you on what to buy the next time you go shopping for food.

Lean Meat

Your main source of proteins should be from the lean meat of naturally raised animals. Being naturally raised means they were grass fed (not grain or corn fed) and are hormone and antibiotic free. According to USDA standards lean cuts of meat should contain only a total of 5 grams of fat, no more than 2 grams of saturated fat, and only 95 milligrams of cholesterol per 3.5 ounce of serving.

They should be trimmed of all visible fat and gristle as much as possible. You can eat as much of these meats as you want and for as long as you follow these cooking tips:

- Make sure you drain off all the excess fat.

- Never deep fry meat to keep added fat to the minimum.

- Instead, try broiling, baking, or roasting.

- If you will sauté any of them try to use as little added oil as possible.

- Limit consumption of animal organs to 85 grams a month.

The following are the recommended lean meat sources and the choice cuts:

- Buffalo or Bison meat (any cut)

- Beef (flank steak, top sirloin, bottom or top round, eye of round, and other lean cuts, ground beef should be 90% lean)

- Chicken (skinless breast meat, extra lean ground chicken meat)

- Turkey (skinless white turkey meat and extra lean ground turkey meat)

- Duck

- Goat (any cut)

- Pork (Bone-in rib chop, bone-in sirloin roast, tenderloin, boneless top loin chop and roast, center loin, ground pork that is 90% lean)

- Lamb (arm, leg, and loin, shank, ground meat must be 90% lean)

- Eggs from non-poultry raised chickens

- Animal organs from the above listed animals like kidneys, bone marrow, liver, tongue, and sweet bread.

- Game meats like rabbit, wild boar, or deer (any cut)

FISH

The micronutrients found in fish have been found to boost brain power and development. Fish should therefore be part of your everyday Paleo diet meals whenever possible. Fish and other seafoods are excellent sources of protein and Omega 3. Remember though to always opt for fish caught in the wild over farmed fish. Avoid canned fish too. Unlike meat, the fattier fish are more nutritious because they contain more beneficial fats and other micronutrients and Vitamins A, C, and E.

The more common wild-caught fish you will find in the market are:

Salmon

Tilapia

Mahi Mahi

Swai

Halibut

Cod

Anchovies

Bass

Sardines

Flounder

Tuna

Other seafood includes Shellfish like shrimp, crab, lobsters, clams, mussels, and scallops.

FRUITS

There are no restrictions for fruits in the Paleo diet. You can eat any fruit you want and consume as much as you want. However, if you are trying to lose weight, it is best to limit consumption of dates, mangoes, bananas, watermelon, and pineapple. They have the highest sugar content and it is best to consume them in moderation. Dried fruits should be avoided as they are packed with sugar. Eat more avocados because they contain a load of healthy fats.

VEGETABLES

All non starchy vegetables like the dark leafy greens, pumpkins, herbs, and seaweeds are allowed in the Paleo diet. Starchy tubers like potatoes and legumes

NUTS AND SEEDS

Raw, unsalted nuts and seeds make good Paleo snacks. The list includes cashews, pecans, macadamia, almonds, pistachios, Brazil nuts, and pine nuts, pumpkin seeds, sunflower seeds, sesame seeds, flaxseeds, and pumpkin seeds.

OILS

Avoid using vegetable oil, peanut oil or canola oil for cooking. They are highly refined oils and as such have higher concentrations of inflammation causing polyunsaturated omega-6 fats than the inflammation reducing omega 3 fats.

Use instead the following:

Virgin coconut oil for all types of cooking

Avocado oil for salad dressing or for low heat cooking

Olive oil for any kind of cooking and as salad dressing too

Sesame oil for seasoning and low heat cooking

Walnut oil for seasoning

Flaxseed oil not for cooking or seasoning but for direct intake as omega 3 supplement

The use of these Paleo recommended oils will help maintain the balance between the omega 3 and omega 6 fatty acids in our body.

Paleo Beverages

Soda and other bottled beverages are out of the list. This includes bottled or canned fruit juices which are normally high in concentrated sugar. Below are the recommended Paleo friendly beverages that must be taken with no sugar added:

Water

Soda Water

Tea

Coconut Water

Coconut Milk

Coffee

Almond Milk

PALEO TREATS and SWEETS

 The Paleo lifestyle is not all that dull and drab. There is room for occasional alcohol as long as you don't overdo it in one drinking session. You may also indulge in dark chocolates to your heart's delight. They are good for the heart. If you are looking for sweetener, use raw honey sold at the farmers market instead.

Chapter 5: Making Sure Your Meat is Paleo

Ideally, Paleo meat must be from free ranging live stocks and are grass fed rather than corn or grain fed. Not only are they leaner, they are also healthier and more delicious. People who have tasted free ranging chicken would swear to this fact.

Meat from herd animals that grazed freely on pasture lands, and pigs and chickens let loose on the field are definitely more delicious as well as nutritious. The first ever staple food of our prehistoric ancestors is meat from wild animals they were able to hunt down.

Paleo is about reversing the ill-effects of excessive industrialization to our lives which has also changed our eating habits dramatically. It is about eating more natural and organic food like the cavemen did millions of years ago and meat was one of their first staple foods.

Like the meat of our ancestors, the meat meant for inclusion in Paleo diets must be from pasture raised livestock. Please take note that there is a big difference between pasture finished livestock and those that has been entirely grass-fed throughout their existence. Most of the pasture finished cows spent the first half of their lives subsisting on grain fed diets and were allowed only to roam the pasture to feed on grass only before they are sent to the slaughter house. They are not entirely free of antibiotics and synthetic food supplements.

Here are some tips to help you make sure that the meat you buy is Paleo friendly:

- Buy only AGA certified grass fed meat. AGA stands for the American Grassfed Association. AGA meat products certification is a guarantee that the product came from livestock fed only with grass from the time they were weaned up to the time they were harvested. The association was formed in 2003 by organic livestock producers, veterinarians, and range management specialists to promote the grassfed industry. They have producer-members in practically every state and their products are sold in supermarkets as well as various farmers markets within the immediate vicinities of their farms. You should not have difficulty finding AGA certified meat products in the country. You may even order meat products online from some of these producers. For more information on AGA producers, visit their website here.

- Other online sources for organic food you may wish to refer to includes:

- Eat Wild

- Local Harvest

- Bay Area Meat CSA (requires sign up for free membership)

- Buy game meat whenever and where ever possible. The best meat that truly approximates the kind of meat eaten by our 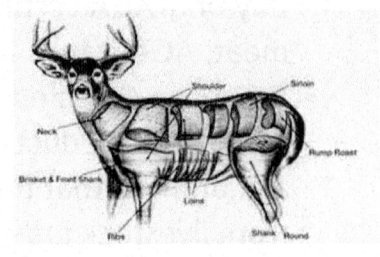 Paleo ancestors is game meat. This includes rabbit, deer, buffalo, ostrich, and wild boar thriving in the wild. They are not difficult to obtain as there are brick and mortar shops as well as online shops where you can buy game meat. They may cost more than the regular meat sold in supermarkets but they are definitely more delicious, more exotic and healthier. There are restaurants too which features exotic game meats in their menus.

- Here are some online shops where you can buy game meat:

- <u>D' Artagnan</u>

- <u>Fossil Farms</u>

- <u>Cavendish Game Birds</u>

- <u>Prairie Harvest</u>

- Don't buy farmed fish and seafood. Farmed fish are those that are farmed in large scale like shrimps, tilapia, salmon, and milk fish. Farmed  fish and sea foods don't get to eat their natural food. They are given fish meal from grains instead. Farmed fish have at times high mercury content. Make sure the fish or sea food you buy is not farmed but caught from the sea, lake, or ocean and not industrially farmed in large scale. How would you know? Ask! Ask the vendor where the fish or seafood came from.

- Don't buy chicken and other poultry products that were raised indoors. Chances are they are grain fed. Look for free ranging chickens. Don't think they are hard to find. You can even find organically grown, free ranging chickens in supermarkets but your best bet would be the farmers markets. There are online sources too and most likely, the same sources where you buy your lean meat also carry free ranging chickens for sale.

Chapter 6: Before You Buy That Meat from the Supermarket

What you should know about Natural Meat and Certified Organic

When it comes to buying meat, especially if it is from a supermarket, it is best that you look at the label first to make sure what you are buying is Paleo meat and comes from livestock that is pasture born and raised and fed only with grass. The tricky part is the meat is allowed by USDA to be labeled either as 'natural' or 'certified organic' and there is quite a significant difference between the two.

 The USDA allows producers to label fresh meat as "Natural" as long as they don't have artificial flavorings or ingredients and uses no preservative. Unfortunately, the USDA does not have any system in place to verify compliance and leaves pretty much everything to the producers to label what is 'natural' and what is not. The bottom line is you can never be sure if the meat labeled "Natural" is indeed free of antibiotics and/or preservatives. Besides, being grass-fed is not a requirement for the meat to be labeled as such. It means that the 'natural' meat you are buying may be free of antibiotics and preservatives, but have been corn fed and therefore not Paleo.

Your best option is to go for meats labeled "Certified Organic". That label means the meat comes from a USDA certified producer whose  breeding history, animal health care system, and type of feed used meet USDA certification requirements. The USDA has inspectors who do on site verification and inspection on a regular basis.

Meat can be "certified organic" if they have been verified to be from animals that have been born and raised in a pasture where they have unrestricted access to the outdoors. Also, they must not have never been given antibiotics or growth hormones during their lifetime.

Here is another tricky part. The USDA considers organic any meat from grain fed livestock as long as the grain is fed to the livestock is certified organic too which means the grain is free from fertilizers and pesticides. But whether the grain is organic or not, it is still grain which means the meat cannot be part of the Paleo diet because it must be strictly grass fed only.

So, the next time you go to the supermarket, look closely at the label. It should not say only 'certified organic'. It must also say 'grass fed'.

Chapter 7: The Paleo Guide to Choosing and Preparing Chicken

Types of Chicken

If you have been living in the suburbs all your life, chances are the only chickens you know are the commercial ones which you usually pick up from the supermarket when you go shopping.

There are actually 3 types of chicken meat – the commercial type, the free range chicken, and the organic chicken.

The Commercial Chickens

The commercial chickens are raised in large scales in overcrowded poultries or cages regularly sprayed with pesticides, fungicides, and herbicides.  They are fed with grain based commercial feeds and their growth is enhanced with hormones and vitamin supplements while their health is fortified with antibiotics to guard against pestilence and disease.

The Free Range Chickens

The free ranging chickens on the other hand, are always loose to freely roam outdoors. They are allowed to feed on plants and insects. They feed freely on their own scrounging for food in their natural habitat and allowed to grow on its own with no growth boosting hormones or vitamin supplements. Basically, they are also organic.

The Organic Chickens

The organic chickens are chickens raised in large numbers in wide but contained and controlled environment where they are allowed to also feed on insects and plants. They are given only natural organic feeds like vegetables but sans grain based commercial feeds. Their environment is pesticide, fungicide, and herbicide free. They are also not given antibiotics or growth boosting hormones.

Chicken's Nutritional Benefits

The chicken meat is low in fat and cholesterol but rich in protein and loaded with vitamins and minerals like niacin which is known to boost the good cholesterol (HDL) in your body, Vitamin B6 which helps convert carbohydrates into fuel, Selenium which is a known antioxidant, and phosphorous which is needed by the body for proper cell functions. Chicken meat is also a good source of protein.

Here is a note of concern for you. If you don't want to get traces of harmful chemicals as well from pesticides, fungicides, herbicides, and from synthetic food supplements and commercial feeds, it will do well for you to stay organic and stick to the Paleo food list.

Preparing the Chicken

There are many ways of cooking chicken. It can be baked or boiled, fried, grilled or broiled, or used for soup or stew. But no matter how you want to cook it, it needs special handling prior to cooking.

Raw chicken meat usually carries bacteria the deadliest of them is salmonella. If the chicken meat is left in the open in temperatures between 40 degrees to 140 degrees, the bacteria will multiply rapidly. Never leave it in the open at room temperature for more than two hours. And, you need to cook it thoroughly to get rid of all the bacteria.

Make sure you keep it refrigerated if you are not going to use it yet. Put it in a sealed container so it won't contaminate other food stuff in your freezer. Make sure you wash your hands thoroughly too after handling raw chicken meat and before you touch other food stuff. Clean and sanitize everything that had come in contact with the raw chicken meat including the sink and other kitchen surfaces.

You can cook and eat beef medium rare or even eat it raw but not chicken. Chicken has to be thoroughly cooked. After cooking you need to check if the inside of the thickest part is no longer pink. If it is, you need to cook it some more to be sure it is fully cooked. You can keep cooked left-over chicken in a refrigerator for no more than two days after which you need to throw it away.

Keeping the Chicken Tender and Juicy

Chicken meat dries up fast. If you want it to remain tender and juicy, it must retain its moisture. You can keep its moisture by soaking it in a brine solution for one hour. To make the brine solutions, simply mix 3 table spoons of rock salt with 4 cups of water.

Different Ways to Season Chicken

Marinating Chicken

Chicken meat dries out fast during cooking. It has to be marinated to keep its tenderness and retain its juice. Marinating chicken meat also gives it flavor. It is an age old technique used to tenderize the meat as well as give it flavor and aroma.

You should always marinate chicken in a refrigerator to prevent any bacteria from multiplying. Boneless and skinless chicken marinate faster than cuts with bones and skin still attached. Marinades usually use acids like lemon juice or vinegar as an added option if there is a need to tenderize the meat. The acid attacks and softens the tough connective fibrous tissues in between the muscles and tenderizes the muscles as well.

Flavor can be anything from sour, spicy, sweet and sour, or smoky and sour depending on the ingredients used. For Paleo cooking, you should only use natural ingredients like organic vegetables, herbs, spices, cooking wine, etc. The longer the chicken is marinated the tenderer and the more flavorful it becomes.

The basic marinade ingredients include vinegar, olive oil, pepper, salt, and herbs. For Paleo cooking, you should check what ingredients are allowed before using them. For example, you should not use refined salt. Instead, you should use natural rock salt.

Using Rubs

For some Paleo chicken recipes, rubs are used instead of marinades. Rubs are a mixture of seasonings like coarse peppers, garlic powder, paprika, and some herbs. They can be dry or in paste form. The paste blend is usually mixed with water, or oil, or horseradish, or other natural liquid seasoning to hold the mixture together. Like the marinade, rubs are used to add flavor to the chicken as well as to tenderize it.

But before you put together your own flavorful rub you need to check if the ingredients you use are allowed in the Paleo diet.

Using Sauces and Basting

Another way of adding flavor to chicken is by using dipping sauces and basting. They not only preserve the moisture but also make the chicken dishes look more attractive. The ingredients commonly used for dipping sauces and bastings include gravy, barbecue sauce, natural meat broth, melted butter, garlic butter sauce and Fresh herbs can be added to the dipping and basting sauces for added flavor and aroma as well as for a more wholesome presentation.

Chapter 8: Paleo Guide to Preparing Seafood

Seafood cannot only make tons of varied sumptuous meals for your dining table they can also load with essential vitamins and minerals as well as poly-unsaturated fats.

Seafood Nutritional Facts

Seafood is one of the best sources of protein, essential trace minerals, and vitamins. It contains mostly poly-unsaturated fats and is low in cholesterol.

The micronutrients found in seafood particularly in fish have been shown to be brain booster and aids in its development. Unlike meat, the fattier fish are more nutritious because it contains more beneficial fats and other micronutrients and Vitamins A, B, B6, B12, C, E, and Biotin and Niacin. Its mineral content includes zinc, phosphorous, and calcium.

Tips in Preparing and Cooking Seafood

Always buy fresh seafood. Frozen seafood should only be an option if fresh seafood is not available. For crabs and lobsters, clams and mussels buy them live.

Make sure you cook the seafood no more than 24 hours from the time you bought them.

For frozen seafood, allow it to thaw inside the refrigerator. Don't let it outside to prevent contamination with bacteria.

When marinating seafood, keep it in the refrigerator.

Use a 3-gallon pot when boiling a 4 pound lobster. Fill the bottom of the pot with up to 1 ½ inches of salt water and put it to a boil. Use a bigger pot appropriate for the size of the lobster to be cooked. Rinse the lobster first in cold water before putting it head first into the pot of boiling water. Cover the pot immediately. The lobster is cooked when the shell turns to a bright red color. The meat should be white and firm.

Cooking time for lobster should be:

10 to 12 minutes for 1lb. to 1 ¼ lb lobster

12 to 18 minutes for 1 ¼ lb to 2 lb lobster

18 to 25 minutes for 2 lb to 3 lb lobster

25 to 40 minutes for 3 lb to 6 lb lobster

40 to 60 minutes for 6 lb to 7 lb lobster

When steaming crabs, follow the same instructions as in cooking lobsters.

You can cook mussels by steaming it for 3 minutes in a covered pot.

Shrimps, clams, and scallops can be sautéed with garlic butter or olive oil and seasoned to taste. You need to watch it to prevent overcooking.

When boiling shrimps, use no more than 4 cups of water per pound of shrimp. Bring the water to a boil first then put in the shrimps and simmer for up to 5 minutes.

When Broiling fish, keep it about 2 to 4 inches above the heat if its body is at most 1 inch thick. If it is thicker than this, place it about 5 to 6 inches above the heat.

Avoid deep frying seafood to minimize added fat. However, if deep frying is unavoidable, make sure the cooking time is only 2 to 3 minutes or until the fish turns golden brown. When sautéing fish, make sure you use only enough oil to cover one side of the fish and cook for no more than 3 to 6 minutes per side.

Chapter 9: Paleo Extreme: Eating Raw Food

People often asked if Paleo is the caveman's diet then shouldn't eating raw food be an integral part of it.

The fact is the Paleo diet encourages the consumption of food in their most natural state as possible. Eating raw food is therefore an essential part of it. But it is not so much because our Paleolithic ancestors ate food that way but mainly because it brings significant nutritional benefits which must not be ignored.

For one thing, raw food is live food. Aside from being rich in nutrients, it also contains natural life energy. Cooking food diminishes this life energy and destroys the nutrients substantially rendering it nutritiously useless. Doesn't it make more sense to put living food into your body than food that has been rendered lifeless by cooking?

If the concept of live food and life energy seem farfetched to you, then at least consider this. Food contains natural enzymes which help break down the nutrients and aid the digestion and absorption of food. Cooking destroys much of these enzymes. Of course, the body has the ability to produce these enzymes to help in the digestion and absorption of the cooked food we eat. But it makes the body work harder to produce enzymes every time the body takes in cooked food and the body can only do so much. On the other hand, eating raw food saves the body from the trouble of producing these enzymes.

Cooking food at 116 degrees Fahrenheit or higher destroys all the natural enzymes and much of the nutrients in them. Eating cooked food with much of the enzymes destroyed due to cooking at high temperature actually forces the body produce these enzymes thus unnecessarily adding stress to it. Raw food diet already contains these enzymes and so it saves the body from this added stress.

Popular raw food diets taken by most raw food converts consist of uncooked and unprocessed fruits and vegetables, nuts and seeds, eggs, sashimi dish from fish, Carpaccio from meat, non pasteurized and non homogenized raw milk, cheese, and yogurt.

Raw food enthusiasts will swear that eating raw food is an exceptional energy booster. They profess that it gives them an instant energy recharge. They also claim they get to sleep better after eating raw food, thus, they need less number of hours of sleep than is normally required. Accordingly, they always wake up feeling full of energy all the time.

These are their claims but the more concrete benefits that can be gained from a raw food diet according to nutritionists include:

- Significant weight loss - since it is basically a low fat, low carb diet
- Better digestion because of the natural enzymes in the diet
- Regularity in the vowel movement since the diet is high in fiber
- Lessens the risk of having cardio vascular diseases and other illnesses linked to trans-fat and saturated fats in the food we eat
- Less water retention because the diet is low in sodium which means it aids in maintaining an ideal weight
- Protects against cancer since it is rich in cancer fighting phytochemicals

Paleo Raw Food for Detoxing

There is another significant benefit that can be gained with a Paleo Raw Food Diet - it rids the body of accumulated toxins. Toxins are harmful wastes or free radicals which are residues resulting from breaking down food. They can accumulate in the body through the years and damage cells which may even lead to cancer. The body is actually unable to completely rid itself of these harmful toxins and through time they accumulate to a point that they hamper cell functions and affect energy production and ultimately lead to the development of diseases like cancer.

Eating fiber rich raw food diet helps the body get rid of these toxins. Raw food diet contains antioxidants that neutralize the cell-damaging free radicals. It actually cleanses our digestive systems and fortifies the immune system.

Chapter 10: Stop Worrying about your Fat Intake

For a long time, people have been made to believe that they should cut down on eating fatty foods or risk having cardio vascular disease. As a result, people developed the mistaken belief that to maintain good health they should avoid fat intake and keep cholesterol levels in the blood to the minimum.

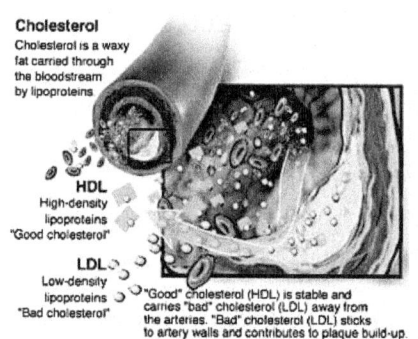

Cholesterol
Cholesterol is a waxy fat carried through the bloodstream by lipoproteins

HDL
High-density lipoproteins "Good cholesterol"

LDL
Low-density lipoproteins "Bad cholesterol"

"Good" cholesterol (HDL) is stable and carries "bad" cholesterol (LDL) away from the arteries. "Bad" cholesterol (LDL) sticks to artery walls and contributes to plaque build-up.

The truth is there are beneficial fats that produce good cholesterols just as there are bad fats that produce cholesterols that are bad for the heart. The secret to good health is in knowing the right kinds of fats to take.

However, this is something you need not worry about if you are on the Paleo diet because everything in the Paleo food list has more of the right kinds of beneficial fats and are therefore beneficial to the heart.

Below is a list of beneficial fats and their Paleo food sources.

Monounsaturated Fats

Monounsaturated fats are known to reduce low density protein (LPL) or bad cholesterol while increasing the high density lipoprotein (HDL) or good cholesterol. HDL tends to stick on the artery walls of the heart causing plaque  to build up which may lead to a heart attack or a stroke. The LDL is more stable and actually transports the HDL away from the arteries thus preventing plaque buildup in the artery walls.

Almonds, cashew, hazel nuts, avocado, red meat, olive and olive oil, canola oil, macadamia nuts, and even poultry products in the Paleo food list are the best sources of monounsaturated fats.

Polyunsaturated Fats

Polyunsaturated fats are known to be cardio protective because of their cholesterol reducing properties. Consumption of omega 3, a form of polyunsaturated fat has been shown to reduce the risk of breast cancer as well as lower the risk of heart attack.

The more common natural sources of polyunsaturated fats include flaxseeds, nuts, leafy green vegetables, fish, algae, sea weed, sesame seeds, sardines, sunflower oil, and wild salmon.

Conjugated Linoleic Acid (CLA)

CLA is known to reduce body fat as well as reduce the risk of colon cancer. Apart from reducing body fat CLA is also known to augment lean muscle mass formation.

Generally recognized as safe by the U.S. FDA, CLA has been added to a number of food categories by different food manufacturers.

Grass-fed cattle produce up to 500% more CLA than grain fed cattle and are therefore one of the best sources of CLA. Among the meat products, Kangaroo meat has been found to have the highest CLA content. Other CLA sources include eggs, mushrooms, and safflower oil.

Medium-Chain Triglyceride (MCT)

Medium Chain Triglycerides or MCT are known to aid the burning of excess fat or calories. Studies made with weight loss subjects who consumed MCT oil have been found to lose more weight as well as fat mass particularly intra-abdominal body fat. They have been shown to help reduce food intake and enhance fat oxidation.

Coconut oil is said to be made up of 66% MCT. Other sources of MCT include Palm Kernel oil and camphor tree drupes.

Fish Oil

Fish oil has been found by researchers from the Linus Pauling Institute at the Oregon State University to be effective as primary prevention of heart disease. The omega 3 fatty acid (EPA) and docosahexaenoic acid (DHA) fish oil is considered as cardio protective and have been known to reduce inflammation all over the human body.

Albacore tuna, shark, swordfish, mackerel and other predatory fish have high concentrations of Omega 3 fatty acids.

Once more, let me point out that it is not the high cholesterol levels in our blood that causes heart disease. Rather, it is the kind of cholesterol running through our blood and the extent of the inflammation in our arteries that causes heart disease.

Critics may argue that the modern urban Western diet has less fat content than the Paleo Diet. Even if this is true, it must be noted that the Western diet is packed with small, highly unstable Low Density Lipoprotein which are easily oxidized. They tend to cling and accumulate on the artery walls and their oxidation is what leads to strokes or heart attacks.

On the other hand, the Paleo diet produces the larger, fluffier LDL kinds 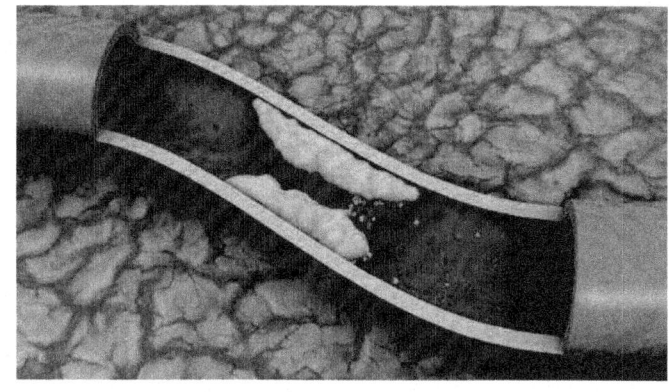 which are more stable and don't oxidize fast. On top of that, the Paleo diet is also packed with HDL (the good cholesterol) which actually carts away the LDL (bad cholesterol) away from the arteries.

Chapter 11: Formulating a Contemporary Paleo Diet Plan

Now that you have more or less a clearer understanding of the Paleo Diet, you have to feel highly motivated to be able to adapt it to your lifestyle. And like any other nutritional program, you have to have a plan to put it into action. You can't just dive in and play it by ear as it will only lead to failure and dismay. Planning your switch to the Paleo lifestyle will guarantee your success as well as allow you to monitor your progress.

There is no "one size fits all" Paleo diet plan. Every plan is uniquely tailored to fit the needs and requirements of individuals. Start with a 28-day plan to eat only Paleo food.

Below are some guidelines you can follow in formulating your planned transition to the Paleo lifestyle.

- Step 1 – Decide on your starting date. Make up your mind when you want the changeover to commence and be firm about it. Don't take too long as you may get sidetracked by other things that will make you lose interest on it. Remember, it is your health and longetivity at stake here.

- Step 2 – Encourage family members to join in changing over to the Paleo diet so that you so won't be all by your lonesome self. Besides, with other members of the family in the same diet, you'd be able to motivate and help each other out. It would be a fun and fulfilling adventure the end results of which is something everyone will cherish for the rest of their lives.
- Step 3 – Make a record of everything before you get started on the program. Take measurements of yourself specifically your height and weight. See your doctor and have blood chemistry done on you to include your blood sugars, CRP, blood pressure, TG, HDL, LDL particle size, etc. All these will help you know the kind of progress you are making with the diet. And oh, don't forget to take a picture of yourself with as little clothes on as possible. This will come in handy when comparing the you then and now.
- Jot down any and all health issues you may have before you start on the diet. Include bloating, diarrhea, constipation, abdominal pain, acid reflux, gas, etc. This should complete the picture of what you are prior to changing over to the Paleo lifestyle.
- Step 4 – Rid your kitchen of all non-Paleo foodstuff. Throw them away or better still, give them out to your neighbors. Whatever you decide to do with them, make sure they are out of your sight and in a place where you can't be tempted to reach for them.

- Step 5 – Go shopping. Get your Paleo food list out and start stocking up on them. Visit your friendly neighborhood grocery store or supermarket and if you can't find some Paleo foodstuff there visit the farmers market. Note down what you can't get from your nearest neighborhood food outlets and order them from online sources. The important thing is to have everything Paleo in your kitchen so that you won't be tempted to use substitutes.
- Step 6 – Start collecting Paleo Recipes and make a weekly meal plan to last the next 28 day challenge. This way, you won't be at a loss on what to cook next.
- Step 7 – Don't take any body weight measurements until after the 28 day period. But carefully note down any physical or emotional changes you may feel in the course of the challenge. This way you won't get disappointed if the changes are slow in coming. Don't worry, the changes will happen.
- Step 8 – No matter what happens or whatever it is you feel stick to your plan until the last day of the 28 day challenge.

At the end of the 28-day challenge, carefully evaluate the progress you've made if any. It is the time to find out if the Paleo diet made you healthier, feel better, lose weight, or if it worsened your condition instead.

Evaluate by:

- Taking another picture of yourself wearing the same clothes when you took the first picture. You can now visually compare if there are differences between then and now.
- Take your body measurements once more and compare your weight and your waist line then and now.
- Get another blood chemistry work done and ask your doctor's help to interpret and compare the results then and now.

There can only be one reason why the challenge will not bring the expected results. It must be because you are not serious enough to strictly pursue the challenge to its completion and somewhere along the line, you've indulged in some exceptions or broke the diet at times.

If you don't notice any improvements for now, perhaps your body needs more time to adapt to the new diet so stick to the diet a little longer until there are manifest results. If you've achieved great progress, then it is up to you to decide whether you should stick to the Paleo diet for good.

Below is a sample weekly Paleo meal plan. You shouldn't have any difficulty in putting together a weekly Paleo meal plan. Start by collecting as many Paleo recipes as you can and chose those that you fancy most. Incorporate them into a weekly meal plan making sure there is variety so you won't get bored eating the same stuff over and over again. There are tons of free online sources for Paleo diet recipes.

Day of the Week	Breakfast	Lunch	Dinner	Snacks
Monday	Almond Flour Muffins	Tuna salad /w apple & orange	Balsamic roast chicken /w cherry tomatoes	Celery sticks /w chicken liver pâté
Tuesday	Green Eggs	Pork chops plus egg salad	Chicken & zucchini hot salad plus dark chocolate mousse	Dark chocolate covered bacon
Wednesday	Bowl of berries & coconut milk	Shrimp salad /w avocado & orange	Cilantro pork stir fry	Spicy pumpkin seeds
Thursday	Almond banana pancakes	chicken & zucchini salad	Beef rib roast /w horseradish & herbs	Veggies /w left-over pork cretons
Friday	Breakfast burritos	Roasted beets & canned sardines salad	Paleo pizza	ginger-snaps
Saturday	Avocado omelet	Salad /w cooked chicken, lettuce & leftover guacamole	Braised beef chuck roast	A handful of almonds
Sunday	Fried eggs /w breakfast pork sausage	Pork stuffed bell peppers	Baked salmon /w asparagus & roasted beets	Piece of fresh fruit roast or dried apricots

Chapter 12: Modern Paleo Principles

The modern Paleo diet is built upon the belief that man's health and well being can be maintained by consuming a diet  consisting of slightly processed food that were available before the advent of agriculture. There are however no set of fixed rules much like commandments written on stone. Everyone is to free to eat as much as he wants as long as what he eats belong to the Paleo food list as described in the previous chapters. There are no zones to follow, neither are there calories to count.

To the more avid Paleo advocates, the diet is simple enough to follow but for the first time converts, the absence of set rules can be confusing and may even tempt them to make excuses for certain exemptions every now and then and break the diet to splurge on certain indulgences.

For the uninitiated, we have listed below some guidelines or tips on what to do and what not to do. They are by no means a set of rules to follow but more of a set of advice to guide you through your first Paleo adventure.

- Wheat, rice, corn, and all other grains have no place in the Paleo diet. They are no better than refined sugar.
- Junk foods and restaurant food prepared with vegetable oil and synthetic food supplements are also out. You should eat real food instead.
- Refined sugar, maple syrup, corn syrup, and artificial sweeteners are definitely out.
- Virgin coconut oil, palm kernel oil, and olive oil together with animal fats such as butter, tallow, ghee, and lard are in. Refined oils such as canola oil, vegetable oil, corn oil, and soy oil are out. Hydrogenated fats are out as well.
- Beans, legumes, peanuts, tubers, dairy products, have no place in the Paleo diet too.
- Paleo is more than a diet. It is a lifestyle which means aside from food, movement in the form of exercises and rest by way of sleeping are just as important. Together, food, sleep, and movement constitute the three foundations of modern Paleo lifestyle.
- Short, high intensity daily exercises will do well for the required movement.

- You should also make sure you get enough sleep every day. Give your body enough time to recover after each work out. Never abuse your body at all.
- Avoid contact with chemicals as much as possible. Don't drink tap water. It is highly chlorinated or fluoridated. Drink mineral water instead. With much the same logic, avoid swimming in chlorinated pools. Use only deodorants that are free of aluminum, fluoride free toothpaste, and more natural organic soap.
- Do some intermittent fasting two or three times a week. You can skip breakfast and morning coffee. You may skip lunch too if you like. By fasting,  you are actually starving the body cells to induce them to reallocate the nutrients from cells that are not functioning optimally for use to fuel other cell functions. You should not do this every day though as the body will learn to adapt to it negating the purpose of fasting.
- You can save a lot if you join CSA (Community-supported agriculture) programs. There must be one in your neighborhood or anywhere near you. They are the cheapest source fresh Paleo food stuff.

- Consider buying food in bulk like buying half a cow, half a bison, or half a lamb instead of buying by the pound. You will save a lot more this way. Besides, you will be able to get all the prime cuts too. You will need a freezer though for this.
- Start learning how to cook shanks, shoulders, tails, trotters, and hocks. You may have been avoiding them in the past in favor of the prime cuts but you will be surprised at how delicious 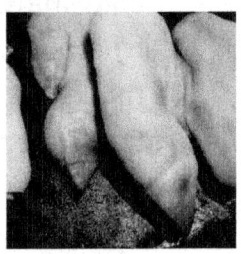 they taste. On top of that they sell cheaper that is why they are called thrift cuts. You will save a bundle of money including them in your meal plan.
- Make it a point to pay the farmer's market a visit near the end of the day and not an hour earlier. The farmers would want to get rid of their produce before the day ends and are likely to sell their produce at a bargain to anyone still hanging around the place rather than leave them to waste.
- Get organized. Make sure you have a Paleo shopping list a weekly Paleo meal plan. Check if all the Paleo ingredients for your weekly meal plan are in stock. Being organizes from shopping to cooking gets things done faster and easier. You will also avoid buying too many things on impulse.

- Make at least one day in a week your "Fish Day." On this day eat only no-cook fish such as tuna salad with kippers for snacks with a lot of raw green leafy vegetables.

"WE ONLY INVENTED COOKING YESTERDAY, AND ALREADY SHE'S SERVING LEFTOVERS!"

- Make recycling very much a part of your cooking. For example, if you grill breast chicken today, don't throw away the left-over. Store them in the refrigerator and use them for chicken salad or chicken stew a day or two later.

- Don't be afraid to experiment cooking your own meal from your very own Paleo recipe. Try to create something that will make you feel great. There are no fast rules about preparing a Paleo meal as long as you stick to using the Paleo food list. Try to discover your own range of Paleo foods that makes you feel great and stick with it.

Once again, don't be discouraged if you don't experience immediate results. There are people whose bodies need more time to adjust to a new diet. And don't be alarmed if at first you feel foggy or lethargic. This is a normal reaction when the body changes over to using fat for energy rather than from carbohydrates.

Chapter 13: There is a Paleo Friendly Restaurant Near you

Cooking your own food is essential to a Paleo lifestyle. However, you should not worry if you are not into cooking your own food or are too busy to do it. There are quite

a number of Paleo friendly restaurants offering natural, organic food in their menus and their lot is increasing every year.

The inclusion of Paleo friendly items in a restaurant's menu would certainly appeal to the growing number of Paleo lifestyle followers as well as others beholden to the concept. The movement for gluten free food has spilled over to restaurants all over the U.S. Increasingly, lean meats from grass fed meats as well as game meats can now be found on the menus of even the trendiest restaurants.

Below is a list of some of these Paleo friendly restaurants arranged in no particular order.

Mashiko Japanese Restaurant and Fully Sustainable Sushi Bar	4725 California SW Seattle WA 98116 Phone.: 206.935.4339	Serves sushi and other Japanese food
Frontera Grill	445 North Clark Street, Chicago, IL 60654 Phone: 312-661-1434 Fax: 312-661-1830	Serves Columbian inspired food.
Topolobampo	445 North Clark Street, Chicago, IL 60654 Phone: 312-661-1434 Fax: 312-661-1830	Sister company of Frontera Grill. Serves Mexican dishes
Namu Gaji	499 Dolores st. 18th San Francisco, CA	This is a Korean Fusion Restaurant. Serves Kimchi and other Korean dishes.
Smoke Restaurant	901 Fort Worth Avenue Dallas, Texas 75208 Phone: 214.393.4141	Serves Paleo inspired omelet with salad inside consisting of tomatoes, asparagus, mushrooms, collard greens, spinach, and your choice of ham, barbecue brisket, barbecue pork, pork-belly bacon, or Andouille sausage.

Crop Bistro and Bar	2537 Lorain Avenue Cleveland, Ohio 44113 Phone: 216-696-CROP	Serves fresh dishes from local producers.
Spin Modern Thai Restaurant	14005 U.S. 183 Suite #1000 Austin, TX 78717 Phone: (512) 258-1365	
Jacks Wife Freda	224 Lafayette St., NY NY 10012 Phone: (212) 510-8550	Try the duck lardon salad, topped with a poached egg.
Meritage	410 St. peter Street, Saint Paul, Minnesota, 55102 Phone: 651-222-5676	Check out the duck fat poached Oregon sturgeon, (make sure to specify without flour)
Neptune Oyster Restaurant	63 SALEM ST BOSTON, MA 02113 Phone: 617.742.3474	Try Hamachi Ceviche in jalapeño-lime vinaigrette w/ cilantro and shaved red onion
Dick's Kitchen	3312 SE Belmont St. Portland, OR Tel: 503-235-0146	Serves only grass fed beef and other Caveman diets.
Calistro California Bistro	6107 N Scottsdale Rd #110 Scottsdale, AZ 85250 Tel: (480) 656-5244	
Juventino	370 5th Avenue Brooklyn, NY 11215	Serves grass-fed beefs and just-picked veggies from the garden
Toro Restaurant	1704 Washington Street Boston, MA 02118	Serves Barcelona-style tapas for Paleo dieters

Juventino	370 5th Avenue Brooklyn, NY 11215	Serves grass-fed beefs and just-picked veggies from the garden
Toro Restaurant	1704 Washington Street Boston, MA 02118	Serves Barcelona-style tapas for Paleo dieters
AUSTIN Foreign & Domestic	306 E 53rd Street Austin, TX 78751	fresh greens paired with succulent clams, short ribs, and pork jowls
The Publican	837 W Fulton Market Chicago, IL 60607	MUST-TRY: Suckling Pig, oysters
Corner Table	2736 Virginia St., Houston, TX	This new restaurant offers a rather large paleo section on Molzan's menu that's been getting rave reviews.
Roots Bistro	507 Westheimer Road, Houston, TX	Naturally paleo-friendly, as it serves raw, vegan and vegetarian dishes with a modern, upscale twist
Sammy's Wild Game Grill	3715 Washington Ave., Houston, TX	Offers plenty of wild game burgers and hot dogs. You can get nearly any wild animal on a salad here.

Cafe TH	2108 Pease, Houston, TX	They have a special Paleo section on their menu.
Georgia's Farm to Market	12171 Katy Freeway, Houston, TX	It is a restaurant with a grocery on the side. It carries a wide selection of grass fed meat.
Snap Kitchen	3600 Kirby Drive, Houston, TX	Offers an entire Paleo section in its stores
Ruggles Green	15903 City Walk, Sugar Land, TX	Serves gluten free, dairy free, and vegetarian dishes
Aladdin Mediterranean Cuisine	912 Westheimer, Houston, TX	Serves steak, chicken and lamb kebabs
Dadami Korean Restaurant	1927 N. Gessner Road, Houston, TX	Serves amazing sushi. The emphasis here is on raw fish and seafood of every stripe
Partners in Paleo	109B Meadow Parkway League City, TX 77573	Serves paleo food only
Lin Jia Asian Kitchen	3437 Lakeshore Ave Oakland, CA 94610 (510) 835-8322	Serves clean meat & produce, no grains, low starches, no refined oils.
Gather	2200 Oxford St Berkeley, CA 94704 (510) 809-0400	Serves gluten/grain-free dishes.
Roam Artisan Burgers	1785 Union St San Francisco, CA 94123 (415) 440-7626	Serves grass fed beef burgers.

Roam Artisan Burgers	1785 Union St San Francisco, CA 94123 (415) 440-7626	Serves grass fed beef burgers.
Tropisueño	75 Yerba Buena Ln San Francisco, CA 94103 (415) 243-0299	Upscale Mexican with a variety of dishes.
Mazza Grill	35 Crescent Dr Pleasant Hill, CA 94523 (925) 680-1858	
Rumi's Kitchen	6112 Roswell Rd NE Atlanta, GA 30328 (404) 477-2100	Serves Persian and Iranian dishes which are basically Paleo
Sabores Del Plata	6200 Buford Hwy Norcross, GA 30071 (678) 743-4671	Serves massive platter of Latin American meat grill dishes.
Farm Burger	410B W Ponce de Leon Ave Decatur, GA 30030 (404) 378-5077 4514B Chamblee Dunwoody Rd Dunwoody, GA 30338 (770) 454-2201	Serves gluten free burgers. Uses only 100% grass fed beef.
Yeah! Burger	1168 Howell Mill Rd Atlanta, GA 30318 (404) 496-4393	
Café Agora	262 E Paces Ferry Rd NE Atlanta, GA 30305 (404) 949-0900	Serves Paleo friendly Turkish and Mediterranean dishes.

Chapter 14: Paleo Diet for Athletes

 It is not only those who have problems with their weight, or those looking for a cure for some ailment or disease who find the Paleo Diet healthfully beneficial. Even the able bodied athletes in the peak of their form have found the Paleo to be the perfect diet for them particularly the endurance athletes who often spend many hours a day in intense training.

For years now, nutritional experts have been trying to look for that perfect food combination that will help boost the athletes' endurance during training and ensure excellent  performance during competitions. The nutritional requirement of the ordinary person who lives a sedentary lifestyle is so much more different from an athlete who is frequently subjected to intense exertions almost daily.

Apparently, the Paleo diet appears to be a God given gift to these champion athletes as most of them would attest to it. With the modern urban diet containing stuff that proves to be prohibitive to competing athletes, nutritionists have been scrambling to put together a diet that is tailor fitted to their specific nutritional needs.

Because athletes undergo intense workouts everyday, eating correctly  before, during, and after every workout as well as in between workouts is of paramount importance to ensure they have enough energy to achieve peak performance and at the same time help their bodies recover fast from the grueling punishment. The main focus of an ideal athlete's diet is to provide him with enough fuel to sustain him through many hours of intense energy output and to provide him nourishment after each session to help the body recover fast.

The ideal diet for endurance athletes must therefore be able to do one thing - to maintain adequate glycemic loads in the blood at all times during the workout. This may mean increasing carbohydrate consumption more often and in larger quantities than the normal requirement of the ordinary individual.

Endurance athletes need to have a moderate intake of carbohydrates two hours before each intense workout. This will give the body sufficient time to bring the blood sugar to the appropriate levels suited for rigorous exertions that requires intense energy output. If the workouts are going to be long winded, the athlete may have to increase his carbohydrate intake occasionally throughout the duration of the exercise. For work-outs that last no more than an hour, drinking water regularly would suffice.

He must also eat foods rich in carbohydrates and with a modest amount of protein within the first half-hour after each workout. This is essential for fast and effective recovery. If necessary, he can continue consuming high carbohydrate foods for several hours for an extended recovery. After which, he must go back to his regular Paleo diet regimen.

Below is a sample outline of a Paleo Diet modified for endurance athletes as culled from the works of various nutritional experts who are experienced in handling endurance athletes. Breakfast, lunch, and dinner were not included in this outline as they are not to be modified and should remain strictly Paleo.

Stage	When to take	What to take	Notes
Before Work out or Race	At least 2 hrs. prior	Low fiber Carbohydrates with moderate glycemic index. Take in another 200 to 300 calories for every hour thereafter.	If carb intake was not possible 2 hrs. before the exercise, take 200 calories of carbs with high glycemic index 10 minutes before actual exercise or race.
During Work out or Race	Throughout the duration of the race or exercise	Eat or drink fluid carbs with high glycemic indexes at 200 calories per hour. This can be modified according to the weight or the duration of the workout or race.	If the work out or race is less than one hour, there is no need for carb intake. Water will do.
After Work out or Race	Within 30 minutes after work out or race	Take a recovery drink that has both carbohydrate and protein with a 5:1 ratio.	Recovery drink recipe you can make on your own: Blend 16 ounces of fruit juice with banana, 3 to 5 teaspoons of honey, two pinches of salt, and 3 tablespoons of protein powder. (You can use 1 egg instead of protein powder.)

For immediate Recovery	For the next several hours and for as long as the work out or the race lasted	Continue consuming carbohydrates with protein with 5:1 ratio.	
For extended Recovery	For the remainder of the day immediately the previous stage	Return to your modified Paleo meals until the next stage 2 hours before the next workout or race.	

The nutrient requirement of an athlete may vary from time to time depending on his training schedule. The carbohydrate and fat intake as shown in the above schedule is designed to go along well with the opposing swings in the athlete's energy utilization as the training progresses from different stages. It is important to note down these energy swings and program the carbohydrate and protein intake accordingly.

The Paleo Diet is Naturally Ergogenic

Perhaps the most significant feature of the Paleo Diet which specifically benefits athletes is it being truly ergogenic. It means it is packed with nutrients that enhance the performance of athletes. It is particularly high in animal protein which is the best source of branched-chain amino acids like leucine, valine, and isoleucine which are basically responsible for muscle growth and repair.

The modern urban diet is commonly acidic and the body's natural reaction to neutralize acidity by breaking down muscle tissues. Athletes on a Western diet normally would find it hard to maintain and beef up their muscle stores after a hard day's workout. On the other hand, athletes on the Paleo Diet do not suffer protein breakdown in their muscle mass. The Paleo diet actually produces a net alkaline effect in the body which is the exact opposite what the contemporary Western diet produces. This eliminates the need for the protein-breaking reaction of the body to acidic diets thus no protein muscle breakdown occurs.

The Paleo Diet for athletes has to be modified though. It must be tailored specifically to an athlete's training regimen and nutrient requirements. It must be modified in such a way as to allow a small window of opportunity for the consumption of starches and simple sugars by the athletes from non Paleo sources during, before, and after the actual exercise. It must allow high carbo intake by athletes as needed and when it suits them best while training both to keep them in their peak forms during the workout session and to give them the much needed nutrients for their bodies to recover rapidly in time for the next grueling exercise.

Outside of the training schedule, it is strictly Paleo all the way – low carb, high protein diet. That means consuming a lot of lean meat, seafood, poultry, vegetables and fresh fruits as much as the athletes like.

To summarize, the Paleo Diet has the following beneficial advantages for athletes as Compared to other diets available today for endurance athletes:

The Paleo diet for athletes has more of the branched chain amino acids (BCAA) which enhances muscle growth and anabolic function. The BCCAs also prevents or minimizes the suppression of the immune system which normally follows intensive workouts by endurance athletes.

- The diet strikes a balance between the inflammation causing omega-6 fatty acids and beneficial omega-3 fatty acids. It reduces if not totally eliminates post-workout tissue inflammations which commonly affect athletes after a strenuous work out.
- The Paleo diet produces an alkaline effect which lowers body's acidity thus reducing the debilitating effect of acidosis on the bones and muscles while at the same time inducing protein synthesis in the muscle tissues for more muscle growth.
- The Paleo Diet includes nutrient rich vegetables and sea foods. They are packed with essential vitamins and trace minerals needed by endurance athletes to maintain optimal health and for effective long term recovery from grueling exercises.

Training for such endurance
sports as running, weight
lifting, cycling, swimming,
triathlon, rowing, and cross
country marathons is truly
taxing to the body. The
endurance athlete is constantly
in some sort of a recovery stage
following every strenuous
workout. This is where the
Paleo diet is vital to an athlete
as it re-nourishes the body with the much needed
carbohydrates and proteins it lost in the training while at
the same time providing the body with trace nutrients
and vitamins needed to repair the body back into form.
Together, the Paleo diet and a truly restful sleep become
the essential components of the endurance athlete's
training program.

The athlete needs to recover fast and effectively after each heavy workout for him to be ready and able for the next. This has always been the biggest challenge an endurance athlete faces in preparation for a competition. Let truth be told, it will be impossible for an athlete to make a full and fast recovery with a strictly all Paleo diet. What he eats should be able replenish the stored nutrients and glycogen he lost on each training session and a low carb high protein diet like the Paleo diet won't suffice as it is.

That is why the Paleo diet has to be modified to include the use of non Paleo high carbohydrate sources during, prior, and post training. It is the only way to ensure rapid recovery in time for the next heavy work out. In short, a modified Paleo diet for athletes has to be formulated to accommodate all the nutrient requirements of an endurance athlete.

Chapter 15: Formulating the Paleo Diet for Athletes

As has been pointed out earlier, the Paleo diet has to be slightly modified to fit the requirements of endurance athletes on training. Specifically, they need to augment their carbohydrate intake at different stages of each heavy workout. Prior to the work out, carbohydrate intake is needed to raise the sugar load in the blood to the appropriate level for the expected intense energy output. High carbohydrate intake during the course of the workout is also necessary to maintain peak performance. And to be able to recover fast from the grueling work out, carbohydrate intake is again needed in the first few hours immediately following the workout to help the body recover fast and be ready for the next.

Again, the modifications should be slight and temporary long enough to last the duration of the training period. There are no fast and firm rules on how one should go about it. It should not however veer away too far from the basic Paleo concepts.

Below are some basic Paleo Principles you can use in customizing your own Paleo diet for athletes:

- Leave out all processed foods from your food list and eat only natural organic foods. Processed foods have synthetic ingredients that may impact your body chemistry and affect your performance as well as your health.
- Make fruits and vegetables, nuts and seeds, as your main source of carbohydrates. Limit your consumption of sugared energy drink supplements to training sessions. Take them only as needed but avoid them for the rest of the day.
- 20 to 25% of your calorie requirement must be from animal protein particularly from grass fed livestock and poultry, game meats and from non-farmed fish and sea food.
- Increase your consumption of fish and walnuts and other sources of omega-3 fatty acids to balance the presence of inflammation causing omega 6 from other food sources.
- Don't eat food fried with trans fat such as canola, corn, and soy oil. If you want to fry food use olive oil or virgin coconut oil instead. Better still, don't fry, grill or broil instead. Processed snack foods are high in trans fats so avoid them too. You also have to limit your consumption of saturated fats.
- Leave out all dairy products from your diet.

- Drink plenty of water. Make it your main fluid intake.

Other more specific modifications on the Paleo diet for athletes will depend on the nature of the sport, the volume and intensity of the training as well as on the physical built of the athlete. Generally, the bigger the athlete is, or the harder or longer the training, the more carbohydrates will be needed - prior to, during, and after the workout.

Power athletes like sprinters, weight lifters, soccer players, and swimmers for example have to consume 1 gram of protein per pound of bodyweight in a day as a general rule. Others have benefited tremendously from what Doctor Mauro Di-Pasquale calls a 'cyclical low carbohydrate diet' where they get to consume higher carbohydrate intake once or twice a week only to condition the body to burn fat for fuel too. Dr. Di-Pasquale is a recognized authority on dietary regimens for athletes.

Chapter 16: Sample Meal Plans for the Endurance Athlete

Here is a sample of a simple meal plan for endurance athletes who are on training. This meal plan came from various online sources and is by no means like a Bible  truth that must be followed to the letter. You can substitute items with other food on the Paleo food list as long as they are what are needed by the athlete for training according to the principles discussed in the previous chapters.

Sample Meal Plan for all athletes starting training before breakfast:

Snack to take before training	A 3 ounce chicken serving taken with half a melon. An alternative snack would be 2 scrambled eggs with one cup blueberries.	
Breakfast after first work out	Grilled salmon taken with hash browns made from sweet potato or shrimp cooked in coconut sauce, chard, broccoli, cabbage	It is important that this meal is taken within 30 minutes after each work out
Lunch	Organic ground beef cooked with marinara sauce with a serving of spaghetti squash baked with garlic butter	
Snack	A hand full of almonds, a medium sized orange, or a can of sardines	
Dinner	Delicious halibut baked fresh and served with a garlic-pistachio sauce and herbs.	

Another example of a Meal Plan for the Power Athletes on a regular but heavy training schedule

Breakfast	4 to 6 egg omelets together with one whole avocado and half a cup of blue berries or strawberries.	
Pre- Work out Snack	4 ounces of grilled chicken and add a handful of macadamia nuts	
Post workout meal	Grilled salmon and stir fried asparagus with garlic, bamboo shoot in coconut milk curry, or mushrooms	
Snack	Take a serving of salmon salad with tomatoes, red onion and avocado, and drizzled with olive oil.	
Dinner	You can have grilled organic rib eye steak or a load of grilled shrimps. Add some mixed green salad with ginger and red onions topped with sesame dressing	

Here is another sample menu for a Twice-A-Day workout regimen for Endurance Athletes.

Pre-workout Breakfast at 5:30 a.m.	Make a healthy blended beverage for yourself. Combine natural green tea (decaf), egg white protein powder, one ripe banana, and almond butter in a food processor and blend well. Take it with baked yam and you will have enough calories to burn during workouts.
Energy drink supplement during the 3 hour trainer bike workout from 6:30 a.m. to 9:30 a.m.	Take carbohydrate gel every 25 minutes.
Post-workout recovery drink at 9:30 a.m.	Make your own home brewed recovery drink for post workout nourishment. Simply combine egg white protein powder, cantaloupe, and glucose and you have a healthy home brewed recovery drink you can take immediately after the work out. Of course should still keep on re-hydrating by drinking plenty of water. But, this home brew will help nurse your body back to form.
Recovery Snack at 10:00 a.m.	Eat raisins. Raisins help neutralize the body's acidity by restoring the body's alkalinity thus preventing further protein breakdown and help you stay in good form for the next workout.
Lunch between 11:30 a.m. and 12:00 noon	Grilled skinless chicken breast, one apple, and sautéed asparagus with drops of flax seed oil.

Another pre-workout snack at 3:00 P.M. before the 2nd workout	Take a carbohydrate rich meal prior to your next workout. You can take chopped egg whites mixed with natural applesauce (unsweetened).
Post workout re-hydrating drink immediately after the 60 minute track workout at 4:30 p.m.	Take carbohydrate gel immediately after workout and drink plenty of water.
Post workout recovery snack at 5:45 p.m.	One banana or other fruits with a high glycemic loads will do. It will help your body recover fast.
Dinner at 6:30 PM	Indulge in a sumptuous Paleo dinner consisting of poached wild salmon placed on top of greens or a plate of steamed kale leaves. You can top your greens with sliced strawberries or avocado and a squeeze of fresh lime juice. Put in a few drops of extra virgin oil if you wish. Garnish with sliced oranges and you have a complete meal that packs energy..

This Sample Menu is for days off from Training for Endurance Athlete

This sample menu was put together by Nell Stephenson, Fitness & Nutritional Professional and Ironman Triathlete. Days off training serve as the long term recovery stage for power athletes and the meal plan is designed to prepare them for the next workout regimen.

Breakfast at 7:00 a.m.	Breakfast consists of whites of hard-boiled eggs, a handful of raw walnuts, steamed broccoli with some fresh blueberries.
Snack at 10:00 a.m.	Raw kale leaves with olive oil, a slice of turkey breast, and one apple
Lunch at 1:00 p.m.	Tossed green salad with lime juice, cucumber, tomatoes and strawberries and topped with olive oil, together with a sliced pear and grilled chicken breast.
Afternoon Meal at 4:00 p.m.	A slice of leftover steak, a tangerine and spinach (raw).
Dinner at 6:00 p.m.	Grilled Salmon topped with lemon and olive oil, some grapes and steamed Bok Choy with ginger.
Evening Snack	A handful of fresh strawberries taken with a cup of herbal tea

Timing the nutrient intake perfectly to jive with specific training activities is of utmost importance. It will ensure that the endurance athlete gets the right amount of nutrients to maintain peak performance at all times as well as provide the body with the nutrients it needs to recover from punishing workouts.

Chapter 17: Paleo Frequently Asked Questions

What is the Paleo diet?

The Paleo Diet is basically a dietary concept founded on the belief that by eating in the like manner our stone-age ancestors did and limiting our food intake to the kind of food available to them 2.5 million years ago, we will become leaner, meaner, and healthier. It is more than just a bunch of well concocted recipes. It is a whole lifestyle which also involves eating contemporary food which is similar to the food they ate and in the most natural state possible.

The Paleo diet is not without scientific basis. It is backed by historical evidence, pure logic and countless studies. The same kind transformative logic is applied to essential lifestyle habits such as sleep and exercise.

What is the underlying logic behind the Paleo diet?

The underlying logic upon which the Paleo Diet hinges based is the assumption that the genetic composition of modern man has been programmed towards the diet man has been eating since the era of our stone-age forefathers and it has not evolved much since then. For more than 2.5 million years, man has had the same natural diet consisting of wild plants and animals so much so that the human genome is believed to be already programmed towards this type of diet and the agricultural and industrial revolution in the last 10,000 years has hardly affected it. This is the transformative logic on which the Paleo Diet is based.

There is proof that our ancestors were perfectly healthy as they can be and are free of the diseases modern man suffer from today. Our genes as well as our physiology evolved from a long process of natural choices which made modern man better suited to eat the food that their genes have evolved with for millions of years.

How is the Paleo diet different from other diets?

First of all, the Paleo diet is not a fad diet. It is sustainable over the long term and promotes overall health and longevity. Unlike other diets, it is not focused on achieving just a single, specific goal such as losing weight, enhancing performance in athletic competition or merely being a part of a disease management program. Unlike those diets, the Paleo diet promotes the over-all health, well being, competitive performance and the ideal weight of individuals by creating hormonal balance in the body.

What is allowed and what is not allowed in the Paleo Diet?

The Paleo diet encourages the consumption of food high in beneficial fats particularly those from animal sources, high in animal protein, and moderate intake of natural carbohydrates coming from fruits and vegetables, nuts and seeds. The beneficial fats should include saturated fats from coconut oil, duck fat, lard, tallow, and butter together with monounsaturated fats from avocados and olive oil.

Why should you be on this diet?

This is a simple, straightforward and easy to follow diet that requires none of the cumbersome calorie counting typical of other diets. It is the original diet humans have been programmed to eat. Following this diet gives you a bevy of benefits that is unmatched by other diets. The benefits include weight loss, muscle gain, better digestion, stronger immune system, slowed aging, more energy, better sleep, less stress, smoother skin, and stronger teeth and bones among many other things. The best part is it protects you from many of the diseases of affluence that has plagued modern man for the last 200 years such as heart disease, diabetes, Crohn's disease

What would be a typical Paleo daily diet?

You can start your typical Paleo day with a simple breakfast of two eggs fried in butter with almond flour muffins and some bacon.

For lunch, you can have Paleo chicken fajitas or Paleo chicken salad summer wrap rolled in a leaf of Romaine lettuce. They are easy to prepare. For specific cooking instructions see our Paleo lunch recipes.

Afternoon snack can be a handful of macadamia nuts or almonds which ever you prefer or you can prepare a bowl of berries mixed with some coconut milk.

Dinner can be a simple dish of sautéed asparagus and mushrooms garnished with minced fresh rosemary springs.

Are bacon and eggs healthy?

Eggs from free ranging chickens feeding on natural diets of plants and insects and bacon from grass fed, pasture born and raised livestock should be free from antibiotics, synthetic food supplements, and growth boosting hormones. They are natural and organic and are therefore healthy.

Organic egg yolks and bacon contain generous quantities of beneficial omega-3 fat and good cholesterol. Both contain healthy saturated fat which is essential for efficient function of almost every cell in our body. Besides, saturated fats should also be our main source of calories more than the carbohydrates.

A modest amount of rock salt in the bacon or egg is actually healthy. As for those who dislike the nitrite in bacon, there are nitrate free bacons available in the supermarkets. Nitrates are naturally occurring compounds and should not be a concern. They can be found in much higher doses in practically all vegetables.

"Cavemen eat this kind of diet because they are more physically active"

Nothing can be further from the truth than this myth floated around by detractors of the Paleo diet. While it is true that the prehistoric caveman was more physically active than the average modern man today, he didn't eat purposely because of the physical activities he engaged in. He ate to survive and whenever he could, he would also find some time to relax, take a nap, and have a good night's sleep.

The principle behind the caveman's diet is simple enough for people to understand and appreciate. When you are highly active and engage regularly in daily exercises, you actually burn more of the sugar reserves called glycogens that are stored in your muscles and liver. They need to be replaced otherwise you won't have the same level of energy for the same activity the next time around. To remedy this and restore your strength and stamina you need more carbohydrate intake. In other words you can afford to have more carbohydrate intake than the average person because of your physically active lifestyle.

Corollary to this, a person who hardly exercises and spends most of his waking hours behind a desk or on a couch watching TV must eat fewer carbohydrates otherwise any excess carbohydrate intake will be converted and stored as fat resulting in weight gain and ultimately obesity which threatens to shorten your existence on earth.

Why is sugar prohibited in the Paleo Diet when it is actually natural?

Sure enough sugar is generally a naturally occurring substance. We find it in almost everything nature produces. However, when it is refined into concentrated granules or powder and used as such like we do today it becomes poison to the human body. It is no longer in its natural diluted form. Continuous consumption of refined sugar can cause spikes in blood sugar levels which may induce strokes or lead to some other health problems like diabetes.

Fruits are the most ideal sources of sugar and carbohydrates because the absorption of sugar from the fruits we eat is tempered and slowed down by its high fiber content. Besides, the vitamins, phytonutrient, and antioxidants in the fruit prevent damage to the cells when sugar is oxidized at the cellular level. They also repair whatever damage may have already been done to the cells because of the oxidation of sugar.

We have grown used to getting our sugar fix by adding teaspoon full of refined sugar to the food we eat and the beverages we drink without realizing we are causing great damage to our own bodies. It is time we get our sugar fix from more natural sources like fruits. It is safer and healthier that way.

Cavemen died young, so why should we adopt their diet?

Again, this is another one of the many myths being floated around by detractors to discredit the Paleo diet. The truth is there is evidence which shows that our Paleolithic ancestors lived longer than what most people thought. Many of them who died young, died in the hands of other predators or by being gored by their own prey while on a hunt. Others died of starvation or accidents, but hardly any of them died of natural cause or illness.

The only reason why modern man lives longer today is because he has access to advanced medical care and technology and not because of his diet. Infectious diseases can now be largely contained and neutralized. Without modern medical facilities people today would die early.

And, on the contrary, there is mounting evidence linking the modern urban diet to such life threatening health conditions such as heart disease, cancer, kidney and liver problems, diabetes, and other diseases of affluence.

How can I stay Paleo even if I eat out often?

If you are on the Paleo diet, you should not have any worries if you have to travel and be on the road for some time. Neither should you worry if you suddenly decide to treat yourself or your family or friends to dinner out of a whim. You can stay Paleo where ever you go. The only problem would be you - how resolute you are in sticking to the Paleo lifestyle whatever you do, where ever you go, and whichever situation you may find yourself in.

In the first place, there are Paleo friendly restaurants scattered all over the country and their number is on the rise. You shouldn't have difficulty finding one to dine in. It is not so much like the old days where you'd rarely find restaurants that serve gluten free and purely organic dishes. Today, gluten free and organic is the 'in' thing among restaurants from the high end classy restaurants to the regular fast food chains and shopping mall food outlets. These Paleo friendly restaurants have been sprouting like mushrooms all over the country and in many major cities all over the world. There are even Paleo food trucks roaming around some cities serving strictly Paleo dishes on the go in street corners and parking lots.

Before you go on a trip or before you dine out with friends and family, make it a habit to check out which Paleo friendly restaurants are located in place where you intend to go. List down these restaurants before you leave the house for the trip or the planned dinner with friends or family. Don't live the house without this list. You can do an online search for Paleo friendly restaurants or check our restaurant list in this eBook.

There may be times you may find yourself in a situation where you get invited by friends to dine out in a non-Paleo friendly restaurant. Don't be shy to ask the waiter if they have gluten free dishes. At the same time feint an excuse why you want your meal to be gluten free like having a serious allergy to grains and meals with traces of grain in them can be fatal to you.

You can be sure they will take this seriously because if there is anything restaurant owners fear most it is having a customer develop a health condition because of their food. Oh, and don't forget to tell them not to use vegetable oil in cooking your meal. Feint the same 'allergy' excuse.

Stick to your Paleo diet even when eating with friends in restaurants or when you get invited for dinner in their homes. Don't feel like the 'odd man out' when you refuse some food. However, when you refuse some of the foods they serve you need to explain why so they won't feel slighted. Just explain that what you refused may do some damage to your health.

You can take this opportunity to discuss with them the Paleo diet and the benefit of eating healthy. Who knows? You may yet win them over to the Paleo lifestyle. And even if they don't convert to Paleo you can at least be sure that the next time they invite you over they will prepare something that is Paleo for you.

Should I go back to a normal diet after I reach my targeted weight?

Many people who go on a diet wants to lose weight so much so that they think that all types of special diet Paleo included is only for losing weight. To them, anything with the word 'diet' has become synonymous with weight loss.

However, the Paleo diet is more than just a weight loss formula. It is not something you should discard after achieving your ideal weight. It is instead a formula for a healthy lifestyle which you need to embrace for the rest of your life if you want a disease-free, healthier body and a longer lifespan.

Besides, it makes no sense to go back to a diet which in the first place is the main reason for your being overweight. Of course, you are free to do as you wish including sliding back to your old eating habits and exposing yourself once more to heart diseases, diabetes, and other modern day afflictions that hounds man today. It's your choice but make it good.

If you have reached your ideal weight on the Paleo diet it means it is effective and there is more reason now to stay on with it. But please take note that weight loss is just part of the beneficial effects of this diet. It will heal the damage done by modern urban diet has done to your gut and still continue to balance the hormones in your digestive system. This will take some time to achieve which means you still need to stay Paleo even long after reaching your ideal weight.

Anyway, if you stay Paleo for some time you'll get used to it and find it not only delicious but also fulfilling.

Shouldn't we eat raw meat the way the cavemen did?

The ideal Paleo diet should include both raw and cooked meat. There is however a sub-category of Paleo adherents who believe that eating only raw Paleo meat is the ideal diet. They call it Raw Paleo. Their number is still small but it is growing. On the other hand, the majority of Paleo advocates prefer to have their meat cooked.

It would be erroneous to say that cavemen always ate their meat raw. Perhaps they did before they discovered how to start fire. But when man learned how to start fire he also learned to cook the food he ate. And there is overwhelming evidence which dates for millions of years to prove he did.

Cooking makes meat more digestible and absorbable by our systems. While it is true that cooking destroys much of the meat's nutrients, it properly compensates for this by making cooked meat easily digestible and highly absorbable. Whatever nutrients are left in the cooked meat, they easily reach every part of the body including the brain, nourishing and speeding up their growth and development in the process.

It is no wonder then why man's brain developed faster than those of the rest of the animal kingdom. Our brain is bigger than those of the other animal species and this is because he learned to cook the food he ate which made the food easier to digest and the nutrients readily available for use by every cell in the human body.

It wouldn't make sense however, to assume that our Paleo ancestors cooked their food as extensively as we do today. At the onset, they must have tried experimenting with it and their diet must have been a mix of cooked and raw meat.

Certainly, there is room for raw meat in the contemporary Paleo diet. In fact the optimal Paleo diet should be a mix of raw and cooked. And if you don't have the stomach for raw meat, you can at least cook it rare or medium-rare. Or, you can try raw fish and homemade sushi for a change.

How much fat, proteins and carbs should I eat?

Unless you are an endurance athlete working out in preparation for an athletic event, there is no counting of calories or measuring of fat, protein, or carbohydrate intake for your Paleo diet. And even if you are an athlete there is no magic number or ratio of food intake which everyone must follow to gain optimum benefit from the Paleo diet. How much fat, protein, and carbohydrates a person needs depends on his individual requirements and personal circumstance.

In case you haven't noticed, the dietary recommendations of each one of those promoting the diet differs slightly from one another. There are hardly any two recommendations that are exactly alike when it comes to recommended dietary intake.

This is because different individuals have different physical built, health condition, personal preferences, and fitness objects. Some simply like to lose weight. Others are more into rigid physical conditioning in readiness for future athletic competitions. There are also those who are into the Paleo diet as part of a health maintenance program to cure autoimmune diseases. Naturally, the nutrient requirement depends on individual requirements which will not be the same as those of the other individuals with different requirements or nutritional needs.

Certainly, the Paleo diet is no magic bullet or a cure-all solution for those who wish to pursue a healthy lifestyle or whatever their objectives maybe. Paleo is not like a 'one size fits all' type of a diet. And, the simplest approach to following the diet is to just eat whatever is natural and organic including a lot of beneficial animal fats.

The safest combination should be high fat, low carbohydrates, and moderate protein. But again, how high is 'high' and how low is 'low' will depend on individual needs and requirements. There are really no rules written on stone for people to follow. You may adopt a particular Paleo diet recommendation but you may have to make certain adjustments to fit your personal preference or taste. But so long as what you eat is natural and organic and belongs to the Paleo food list, it will still be Paleo.

The Paleo diet for endurance athletes is a different story especially if the athlete is on to a rigorous training regimen. According to physical fitness coaches an athlete will need to consume 200 to 300 calories from easy to digest carbohydrate sources an hour before his scheduled workout with another 200 to 300 calories every hour thereafter for the duration of his work out. He also has to take another 200 to 300 calories within 30 minutes after the workout to help the body recover from the strenuous workout.

Are supplements allowed?

Doctors would normally recommend supplements to persons with specific health conditions consistent with severe vitamin deficiencies otherwise they would instead recommend a diet rich in the deficient vitamins or minerals.

The Paleo diet is already dense in vitamins and minerals which the body needs. Unless the deficiency is serious enough to require 'shock treatment supplements are totally unnecessary. Besides, most supplements are synthetically produced. Or, they may come from organic extracts but they still contain synthetic ingredients. They can do more harm than good to your body in the long term.

If you are on the Paleo diet there is no more need for supplements. It will just be a waste of your money since the diet has all the nutrients you need.

All you need to do is stay with the diet and let it heal your body naturally – without supplements. Give it sufficient time to work your body and heal it from years of damage caused by the food you've been eating ate. In time your gut will finally have the ideal hormonal balance it needs to function efficiently.

If you are from a northern county or work indoors most of the time chances are you are vitamin D deficient. Go out under the sun more often and for longer periods. You'll get all the vitamin D you need from sunlight. What is more it is free.

Is there a break-in period to the Paleo diet?

Weaning over to the Paleo diet from your old diet you will need may require an adjustment period of about 3 to 4 weeks. It is no joke to make a sudden shift to a low carb diet when your body is so used to being bombarded with high carb doses everyday for many years. You may encounter some dizziness or feel lighthearted most of the time. You may even be edgy and irritable. These are normal bodily reactions if you start taking less than 50 grams of carbohydrates which is all that the Paleo diet will give you every single day. But don't worry because these withdrawal symptoms will soon blow over. And as you get used to the diet more and more each day, you will start to feel highly energized.

The Paleo diet has a detoxing effect on your body. It starts to get rid of the toxins that have accumulated in your muscles cells all these years. These toxins are released to the blood stream and ultimately excreted out of your body by your system. You will be experiencing detoxing symptoms such as dizziness or irritability during this stage but this is only temporary. You will only experience such symptoms while the diet is still cleansing your body after that you will feel energetic and active.

How about withdrawal symptoms?

You may experience caffeine withdrawal headaches or the so called 'low carb flu' manifested in ways like feeling weak, feeling tired, having headaches, or other similarly light symptoms. You need not worry because they will disappear in a few days. Just remain resolute and stick to the diet. Everyone who weaned over to the Paleo diet always come out grateful because the discomfort they went through for a short period of time was nothing compared to the immeasurable benefits they gained from it.

How long will it be before I start feeling better on a Paleo diet?

After 14 to 30 days on the diet the average person should start feeling the full beneficial effects of going Paleo. Of course this may vary from person to person depending on how each one strictly follows the diet. If you go strict Paleo right from day one you may experience more intense but shorter detoxing syndromes. But, you will feel much better from the fourteenth day onwards.

Aren't fish and sea foods dangerous because of high levels of mercury in them?

Fish or any food for that matter which has high levels of mercury is hazardous to your health. This is especially true for industrially produced farmed fish or those that have been raised in fish pens or cages. However, the truth is fish caught wild from the ocean has less mercury content than the vegetables we buy from the supermarket. Fish should be an integral part of the Paleo Diet because it is the best source of high levels of Omega 3. That is why the next time you buy fish you should check the labels to make sure it is not farmed fish but fish caught from the ocean. The better option is to buy fish from the fisherman's wharf if you live near one.

Is the Paleo diet a mere fad diet?

The Paleo diet is more than a programmed food intake. It is a total lifestyle that revolves around eating only natural and organic food stuff. It can't be a fad since this kind of diet has been in existence for millions of years. It is definitely sustainable for a lifetime and not something people can easily lose interest on after some time.

Isn't this just like Atkins?

Many people mistakenly think so. On the onset, they do look similar but if you look at them closer you'll discover that there are significant differences between the two that sets them apart. For one thing the Paleo diet allows you to eat as much fresh fruits as you like – no restrictions while Atkins only allows controlled portions and not all fresh fruits are allowed.

But the most important difference between the two is the fact that the Atkins diet allows consumption of a lot of processed meats and saturated fats. The Paleo diet on the other hand bans all processed meats and puts emphasis on animal protein and beneficial fats from grass fed, free ranging livestock only.

Can the Paleo Diet really help you lose weight?

The simple answer is yes. Weight loss is but a positive residual effect of the Paleo diet. Its real value is in keeping you healthy, fit, and feeling so much better for the long haul. Do not expect it to be quick fix for weight gain. You will lose weight alright but it won't be so dramatic. Rather, it will be a gradual process which makes it highly manageable and truly sustainable and more permanent. With the Paleo diet the body is taught to use fat for energy and limit its dependence on carbohydrates for energy. In the process there will hardly be any excess fat to be stored and whatever fat is there will be converted into energy to power cell functions. You also get to avoid high carbohydrate intake which can cause spikes in blood sugar levels.

The bottom line is majority of people who adopted the Paleo diet achieved significant weight loss after being in the diet for some time. The good news is it permanently corrects the body's dependence on carbohydrates and you'd be able to maintain your ideal weight with hardly any effort.

What can I get from following the Paleo Diet?

Obviously, weight loss is one of the more significant benefits you can gain from following the diet. There are tons of other benefits it provides like getting to sleep much better every night, improved skin complexion, improved sex drive (if you want to consider it significant), improved digestion, clearer mind, and increased energy. The Paleo diet heals your body back to its original condition much like overhauling a car and repairing damaged parts. It can cure such diseases as Crohn's disease and other autoimmune diseases.

Won't you get bored eating almost the same thing evryday?

Theoretically, the answer would be yes you may get bored. The Paleo diet do not have an extensive line of sumptuous recipes as what you have been used to for years and may definitely taste differently with sugar and salt intentionally left out. But if you consider the fact that our body looks for variety only because it is not getting optimal nutrition from the food you eat then you'll better appreciate and understand why Paleo converts are sticking to the diet for good.

Your choices of food on the Paleo diet may be limited but they give you optimal nutrition which ultimately curves your cravings for variety. Besides, it is not true that the Paleo lacks variety. The Paleo recipes included in this book are just a few of the many Paleo dishes you can find around. And they are by no means less sumptuous.

People tend to think of the Paleo diet as another one of those bland and tasteless slimming diets. Nothing can be further from the truth and the thousands of recipes you will discover online is proof of this. You can have a variety of healthy foods without venturing out of the Paleo guidelines. While many choose to stick to a simple diet, you can elect to have variety in your own Paleo meals and a simple online search will help you do that with ease.

Why should I avoid adding salt to my food?

Intake of salt as well as grains, legumes, and cheese create a highly acidic environment in our bodies and put a tremendous stress on our kidneys. As an autoimmune response, the body is forced to tap the calcium reserves in our bones to neutralize the acidity and restore balance once more. The net effect of this however leads to degenerative diseases like osteoporosis.

Grains have fibers, minerals, and vitamins so why should I take it out from my diet?

The little known fact about whole grains is that they contain an indigestible substance called phytate which stores energy and the element phosphorous in the grain. Phytate is the salt form of phytic acid which binds the magnesium, iron, zinc and calcium in our intestines and effectively leading them out of our bodies preventing them from being absorbed by our system. Mammals including humans cannot digest phytate because we don't have the enzyme phytase. Without which it can't be digested or broken down leaving it to wreck havoc by blocking the much needed nutrients we mentioned above from being absorbed by the body. This in effect makes grains anti nutrients.

We do need fiber in our diet fruits and vegetables can provide the fibers we need. Fruits and vegetables are healthier sources of fibers than grain which do more damage than good to our bodies. Unlike the fibers found in grains, the fiber in fruits and vegetables are highly soluble and are easily assimilated into our system without doing damage to the walls of the intestines which grain fibers notoriously do. On top of that, fruits and vegetables contain more B vitamins and folate than grain.

It doesn't make sense to continue consuming grain which has less nutrients and does more damage to our system when we can get more nutrients from healthy sources such as fruits and vegetables without risking damage to our intestines.

Can a vegetarian be on Paleo Diet at the same time?

The simple answer to this is no. They have to choose between the two because of fundamental difference which prevents them from mixing together like water and oil. For example, most vegetarians depend heavily on the use of grains and legumes like beans, peas, as their main source of carbohydrates for their daily calorie requirements. As you well know by now grains and legumes are not allowed in the Paleo diet. They are anti-nutrients and cause chronic inflammation of the intestines a condition called leaky gut which is a precursor to many autoimmune and heart diseases as well as cancer.

Besides that a strictly vegetarian diet deprives the body of the much needed nutrients, vitamins, and minerals that are normally found packed in animal food sources. The vegetarian diet specifically leads to deficiencies in vitamins D, B6, and B12 and essential minerals like iron, zinc, and iodine and beneficial nutrients like omega 3 fatty acids and taurine.

Is the paleo diet going to punch a hole in my pocket?

It is a fact that processed foods produced in large quantities are cheaper than organically grown fresh food. Without a doubt, going strictly Paleo is going to cost you more. However, if you think of the longer term is going to free you from the travails of the modern urban diet and save you from future medical expenses you are likely to incur once you get afflicted with the many diseases resulting from prolonged consumption of processed food. In the end you are likely to spend more in medical bills than the money you saved by buying cheaper processed foods.

Is it practical to bring a Paleo pack lunch to work?

Why not? A pack lunch consisting of vegetable salad topped with grilled chicken breast is easy to prepare. Add an apple or a bag of nuts and pack some vegetables and you have a perfect lunch to go. And if you don't have the time to prepare a meal yourself you can try the prepackaged Paleo meals which are now available from several online sources and in several choices. Just reheat it in a microwave oven at your workplace and you have a hot, healthy lunch.

Do I have to go to specialty food store to buy my Paleo food stuff?

Natural organic foods are not only available in specialty shops like Trader Joe's. They can be found even in your neighborhood supermarket. You just need to read the labels carefully. If you want to really be sure what you are getting is natural and organic then make it a habit to visit the farmer's market near you. You'll get them fresh there. For organic meat and poultry, the best way is to get them direct from the certified producers. You can do this by join any CSA (Community Supported Agriculture) organization. I am sure there is one near you. Not only can you be sure that you will you get your meat and poultry from organically raised and grass fed livestock, you'd also be able to buy them fresh and at remarkably cheaper prices.

Part 3: Contemporary Paleo Diet Recipes

Paleo Breakfast Recipes

Muffins made from Almond Flour

Ingredients:

1 cup blanched almond flour (4 oz)

2 large eggs

1 tbsp agave nectar or honey

¼ tsp baking soda

½ tsp apple cider vinegar

Instructions:

Combine almond flour and baking soda in a medium size bowl.

Mix the eggs, agave and vinegar in a larger bowl

Cut in the dry ingredients into egg mixture, mixing well until the texture is consistent.

Scoop batter into a paper lined muffin pan ¼ cup at a time.

Heat oven to 350° and bake muffins for 15 minutes or until the edges are slightly browned

Cool the muffins while still in the pan for at least half an hour.

Add butter and raspberry jam

Makes 4 muffins.

You can use the same ingredients and the same instructions to make a quick bread loaf. Just double the ingredients and bake it longer – 35 to 40 minutes instead of 15 minutes. Use a small (6-½ x 4-inch) loaf pan.

Scrambled Eggs with Kale Leaves

Ingredients:

4 eggs

4 large untrimmed kale leaves

A pinch of kosher salt

oil for the frying pan (make sure it's Paleo)

Instructions:

Place all ingredients in a blender.

Blend until smooth on high setting.

Place oil in pan on medium heat.

Once the oil is hot enough pour in the egg mixture.

Allow the egg mixture to cook a little and then scramble.

Cook to your desired doneness.

Serve

Paleo Nutty Bars for Breakfast

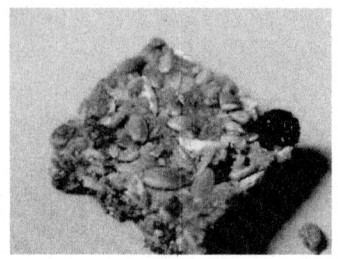

Ingredients:

1 C almond flour (blanched)

¼ tsp kosher salt

2 tbsp honey

¼ C coconut oil

1 tbsp water

½ C shredded coconut (unsweetened)

¼ C raisins

½ C pumpkin seeds

¼ C almonds (blanched slivered

1 tsp vanilla extract

½ C sunflower seeds

Instructions:

Combine and mix salt and almond flour in a food processor.

Cut in water, honey, coconut oil, and vanilla.

Add shredded coconut, almond slivers, raisins, pumpkin seeds, and sunflower seeds.

Place the dough firmly into a standard baking pan (8 x 8 inch), and pat down the dough using your hands wet with water.

Bake for twenty minutes at 350°.

Boneless, Shredded Pork Rib Tacos for Breakfast

Ingredients:

Tortilla Ingredients

½ cup coconut milk

3 eggs, whisked

Two (2) tbsp coconut flour

Salt to taste

Shredded pork Ingredients

2 lbs pork short ribs (boneless)

Two (2) tbsp maple syrup

Two (2) tsp garlic powder

Pinch of salt

Toppings

5 to 6 strips of bacon

6 to 8 oz of green chilies (1 can)

green onions, chopped

2 to 3 tbsp hot sauce

Instructions:

Put short ribs to the Crockpot.

Pour in maple syrup together with garlic powder and salt. Make sure the shorts ribs are well coated.

Cook for 8 to 10 hours on low heat.

Remove the pork ribs and shred them. Pour some of the excess juice over the shredded pork for added flavor.

To make the tortillas,

Use a medium bowl and whisk together all the ingredients until smooth.

Use a non stick skillet and set it on the stove on medium-high heat.

Pour enough mixture into the pan to make a pancake size tortilla. You can pour on the mixture on the skillet and roll it to spread it around evenly. Flatten out the tortilla some more if necessary.

Cook each side for one minute.

Cook bacon on the skillet and chop into roughly pieces or bits..

Mix the green chilies with the hot sauce. Place in a bowl and microwave for 1 ½ to 2 minutes or until hot.

Cool and then chop the green chilies.

Load each tortilla with shredded pork, bacon, green onions, and green chilies and serve.

Chocolate Donuts with Coffee and Bananas

Ingredients:

3 ripe bananas

3 eggs

5 dates, pitted

¼ C coconut flour

2 tbsp coconut oil, melted

⅛ C honey

1 tsp vanilla extract

½ C chocolate chips, dark, melted

¼ tsp baking soda

1 tsp cinnamon

¼ tsp baking powder

pinch of salt

1 tsp ground coffee

Sea salt

Instructions:

Blend bananas, dates, and honey, in a food blender until smooth.

Add eggs, oil, cinnamon, coconut flour, salt, baking soda, and baking powder and pulse until you have a complete puree.

Grease a donut pan and pour in the pureed ingredients.

Place inside an oven preheated to 3750 F. Bake until completely cooked (about 20-25 minutes).

 Test if cooked by Poking toothpick through one. If it comes out clean then it is done otherwise bake a little longer.

Allow the donuts to cool.

Use another bowl to melt the dark chocolate. Stir in the ground coffee.

Dip each donut (top half only) in the coffee and melted chocolate mixture.

If you wish may sprinkle top with sea salt.

Meatza for Breakfast

Ingredients:

1lb Bison Breakfast Sausage

7 eggs

6 to 8 slices of bacon (diced)

½ sweet potato or yam (diced into smaller pieces)

½ diced yellow onion

1 minced garlic clove

Instructions:

Put the breakfast sausage in a medium size bowl and crack one egg directly into it. Mix using your hands to break up the egg and continue until well combined.

Pour the meat mixture into a 8×8 glass baking dish. Try to even the surface of the meat mixture by pressing down until the whole surface is even.

Put the baking dish in an oven that was pre-heated to 350 degrees. Bake for 8-10 minutes only or until fat rise up to the top.

Take the baking dish out of the oven and drain the excess fat.

While the meatza is oven, cut up the bacon and cook in a large pan on medium heat. Cook until crispy. Remove the bacon from the pan and place on a plate with a paper towel to drain the excess fat. Pour the excess bacon fat in the pan into a container but leave 2 to 3 tablespoons of the excess bacon fat in the pan.

Add your garlic clove, your sweet potato and yellow onion in that order to the pan.

Let the onion and sweet potato simmer for about 8 to 10 minutes. Stir frequently to prevent it from burning.

Once the sweet potatoes are soft and your meatza is done, you can now start making the layers.

Top the meatza with the sweet potato/onion mixture making sure it is evenly distributed.

Crack 6 eggs directly on top of it.

Top with bacon and place it back inside the oven.

Cook for another 8 to 10 minutes or until eggs are cooked to your liking. If you wish you can make scrambled eggs instead and use it your topping for the meatza.!

Let cool and add hot sauce before serving.

Prosciutto Crudo with Baked Eggs

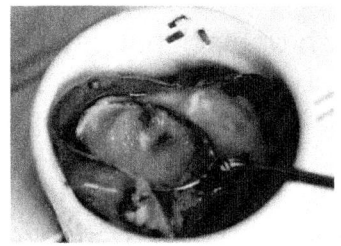

Ingredients:

1 slice of prosciutto crudo

1 pasture-raised chicken egg

chives

fresh ground black pepper

1 tsp heavy cream (optional)

Instructions:

Preheat oven to 375 degrees F.

Line one bouillon bowl or ramekin with one slice of prosciutto. Never mind if it overlaps on the edges.

Crack one egg directly into bowl lined with prosciutto.

Add one teaspoon of heavy cream.

Spice up with a dash of black pepper and garnish with a sprinkle of chopped chives.

Place the bowl on a baking pan and place inside the oven which was pre-heated to 375 degrees.

Bake for 15 minutes.

Take the bowl out of the oven and let stand for 5 minutes before serving.

Paleo Porridge

Ingredients:

2 bananas (ripe, mashed)

2 C coconut milk

1/4 C flax meal

3/4 C almond meal

1 tsp cinnamon

1/8 tsp ground cloves

1/2 tsp ginger

1/8 tsp ground nutmeg

maple syrup or raw honey as sweetener

1/8 tsp coarse sea salt

You may add unsweetened coconut flakes, berries, nuts, or seeds, for topping.

Instructions:

Use a medium size saucepan. Put in all the ingredients and simmer on low heat.

Stir occasionally until thick and produces bubbles. It will be initially thin but will gradually thicken as you cook.

It will thicken some more after cooking till up to serving time so it is best that you add a little extra coconut milk or water.

Top with berries or coconut flakes before serving.

Kale Salad

Ingredients:

1 handful of kale leaves

3 ounces Andouille sausage

1/2 cup sliced mushrooms

½ cup Vidalia onion, diced

1/4 cup + 1 tablespoon extra virgin olive oil

2 tbsp apple cider vinegar

1 tsp + more to taste

1/2 tsp cracked black pepper

2 eggs fried to "over medium"

Instructions:

Trim the kale leaves by removing the stems and thick spines.

Wash, and pat dry thoroughly. Chop or hand tear the kale leaves into bite-sized pieces, set aside in a large mixing.

Dice the Andouille sausage and saute' in medium fire using 1 tablespoon of the olive oil and a medium size frying pan.

Cook sausage until much of the fat has been rendered and the sausage is somewhat crispy. (about 5 to 7 minutes).

Add the mushrooms and onions and cook for another 5 minutes.

Turn the fire to low and add the 1/4 cup of olive oil together with pepper, salt, and the apple cider vinegar.

Stir well for a few seconds until everything is heated through.

Pour the still hot Andouille sausage dressing over the kale and set aside.

Fry two eggs "over-medium".

Toss the salad with the dressing and divide between two plates.

Serve with either the egg alongside the salad or with the egg on top of it.

Paleo Lunch Recipes

Beef Barbacoa

6 pounds beef roast (trimmed and cut to at least 4 pieces)

8 cloves garlic

6 pieces chipotle peppers (if using canned choose one in adobo sauce) 6 peppers)

1 tbsp dried oregano

3 tbsp oil (ideally coconut, lard, ghee, tallow)

2 tbsp ground cumin

1 tbsp black pepper, ground

5 bay leaves

1 tbsp Kosher salt

1/2 cup of apple cider vinegar

6 tbsp lime juice

1/4 tsp cloves, ground

1 1/2 cups chicken or beef stock

2 pieces juniper berries

Instructions:

Heat two tbsp of oil with medium to high heat using a large pan.

Pan Fry each side of the meat for about 2 to3 minutes making sure they are seared or properly browned.

Place the seared meat in a Crockpot.

Put together the vinegar, adobo sauce, lime juice, oregano, chipotle peppers, garlic, cumin, salt, black pepper, and cloves in a food processor and blend until smooth.

Pour blended sauce over meat and then add the beef stock, juniper berries, and bay leaves.

Cover the pan and cook for 6 hours on low heat or until the meat is very tender.

Transfer the liquid into a wide bottomed pan and let simmer on high heat until volume is reduced by half.

Chicken Liver Salad with Raspberry Vinaigrette

Ingredients:

10 ounces chicken livers

2 tbsp olive oil

8 cups of arugula

Kosher salt

black pepper, fresh, cracked

2 tbsp balsamic vinegar (raspberry)

Instructions:

Set the arugula leaves in a salad bowl.

Coat a skillet with oil and heat. Simmer the livers in it for about 8 minutes. Turn and squish them every now so that they are evenly cooked.

Remove the pan from the fire when the livers are done. Put vinegar on the pan to take away the glaze.

Let the vinegar and oil mix together.

Sprinkle the liquid from the pan to the salad and toss well.

You may season with salt and/or pepper if you wish.

Divide the salad into two plates and top each with half of the livers.

Serve warm.

Steak & eggs

Ingredients:

1 large steak (It can be filet, sirloin, rib eye, etc.)

2 to3 tablespoons cooking oil (it can be tallow, ghee or butter)

2 eggs from free range chickens

Paprika to taste

Salt to taste

Pepper to Taste

Instructions:

Leave your steak standing for a while at 680 F (average temperature of a room) for about 40 minutes.

Put the pan on medium to high fire and put 2 tablespoons of cooking oil.

Rub the steak with black pepper and salt and before putting in to the pan.

Broil your steak to preferred doneness.

If you wish your steak to be medium rare broil each side for 3 minutes for a perfect doneness.

Once done, take it out from the pan and set aside.

Lower the heat settings between medium and low.

Pour in the remaining cooking oil.

Crack the eggs directly into the frying pan.

Sprinkle with pepper, salt and paprika, according to your preference. Cover and simmer till the egg whites are just right.

You can top the steaks with eggs or serve the steak with eggs on the side.

Paleo Chicken Fajitas

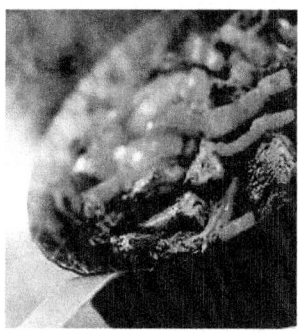

Ingredients:

3 pounds chicken breast, sliced thinly;

3 onion bulbs, sliced;

3 pieces of bell peppers;

6 chopped garlic cloves;

Juice of 5 lemons;

2 tablespoons each of oregano, cumin, coriander and chili powder

4 tablespoons of coconut oil (ghee and tallow are perfect for this too)

lettuce leaves and butter for serving the fajitas;

You may choose the toppings for your fajitas. It can be diced tomatoes, sliced avocados, fermented pickles, guacamole, sauerkraut, and mayonnaise or salsa verde.

Instructions:

Mix together the chicken, onions, bell peppers, garlic, lemon juice and spices, and in a large bowl. Toss well.

Put the mixture inside the refrigerator to marinate for 4 hours.

After 4 hours, heat a skillet large enough for the chicken over medium fire and add your cooking oil.

Place the marinated chicken together with the marinade into the hot skillet and cook until the onion and bell pepper are tender and the chicken is thoroughly cooked.

Transfer the chicken mixture from the skillet to a bowl large enough to accommodate all of it.

People can put together their own fajitas or you can do it for them. To make a fajita simply put enough of the chicken mixture on the lettuce leaves and garnish with any topping you like.

Baked salmon and roasted beets with asparagus

Ingredients:

4 salmon fillets (fresh);

4 tablespoons butter (or you can use coconut oil);

4 teaspoons chopped dill;

16 sprigs asparagus (remove the base)

4 red beets, medium size cubed

Salt

Pepper

Instructions:

Prepare four pieces of aluminum foil and arrange beet cubes and 4 asparagus on each foil.

Set one salmon fillet on top of the asparagus and beet in every foil.

Top each fillet with one tablespoon butter and one teaspoon chopped dill.

Close each foil with the sides touching each other like a pocket.

The baking time for the fish fillet should be 10 minutes for every one inch thickness. If the fillet is two inches thick your baking time should be 20 minutes.

Examine the fish regularly so you won't overcook it.

Top with fresh herbs like dill or cilantro and serve.

Garlic mussels with White Wine

4 lbs fresh mussels;

2 chopped onions;

5 finely chopped cloves garlic;

2 cups white wine or chicken stock;

6 tbsp butter or ghee;

1/3 cup of your favorite chopped fresh herbs (parsley and basil are excellent);

Instructions:

Wash the mussels thoroughly and remove the beards.

Throw away all of the opened mussels cooking cooking.

Use a stockpot for cooking. Place the white wine, garlic and onions and bring to a boil.

Simmer for about 5 minutes before adding the cleaned mussels.

Cover and raise the temperature to medium high. Let it boil until all the mussels are open.

Add the herbs and butter (or ghee) before removing the pot from the fire.

Serve in bowls with butter sauce, white wine, and garlic.

Pork chops with apples and onions

4 pork chops bone-in and with trimmings;

2 large onion, sliced;

3 tbsp lard, (you can also use animal fat or ghee, raw butter, or coconut oil;

4 apples, sliced, with core removed;

pepper

Salt

Instructions:

Rub the pork chops with salt and pepper.

Place a large pan on a stove over medium-high heat and melt two tablespoons of your preferred cooking lard.

Fry each side of the pork chops for five minutes or until browned and properly cooked.

Set aside for a while.

Set the heat between medium and low then add the remaining 1 tablespoon of lard.

Add in apple slices and sliced onions.

Cook for at least four minutes or until the apple slices are a bit soft and the onions appear to be already caramelized.

Top the pork chops with cooked apple slices and onions and serve.

Gluten-Free Chicken Strips

1 1/2 lbs. chicken breast halves, (skinless, boneless, 1-inch strips)

1 tbsp grape seed oil

⅓ C coconut oil (extra-virgin)

½ cup coconut flour

¾ cup shredded coconut

½ tsp salt

12 grinds fresh black pepper

1 egg

Instructions:

Set the oven temperature at 4000 F. While preheating the oven, prepare the flour mixture.

Combine flour with pepper and salt in a mixing bowl (medium size).

Beat the egg together with 1 tbsp of grape seed oil in a separate bowl.

Place the shredded coconut in another bowl.

Dip the chicken pieces first in the flour mixture making sure each piece is coated evenly.

Dip the chicken pieces into the egg mixture then on the shredded coconut.

Arrange the coated chicken pieces in a low sided baking pan and drizzle with coconut oil or melted butter.

Bake for fifteen to twenty minutes turning the chicken over once.

Simple Shrimp Scampi

1 lb of large shrimp or prawns (de-veined but with the tail still on)

3 tbsp pasture butter or ghee

3 cloves of garlic (finely chopped)

1/2 lemon juiced

Salt and Pepper

Instructions:

Heat your frying pan first before adding the butter.

Once the pan is hot enough reduce the heat to medium low and add the butter.

Add the garlic once the butter is melted.

Add the previously washed and dried shrimps.

Sauté the shrimps for 10 minutes or until they are no longer translucent.

Remove from the fire and add the lemon juice.

Whisked with some salt and pepper before serving hot.

Tuna Salad

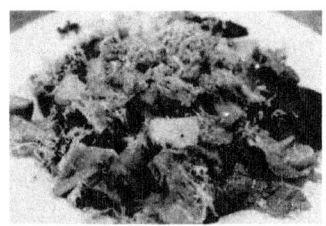

Ingredients:

1 can wild caught tuna in water or oil

1 heaping handful of mixed greens, spinach, mesclum,

1 large boiled beet

1/2 apple diced

2 sprigs green onion, chopped

1 large carrot, sliced

Sprinkle of shredded raw,

2 Tbsp dressing of your choice

Salt and Pepper

Instructions:

Combine all the ingredients except the tuna cheese, dressing and salt and pepper in a bowl.

Open the can of tuna and drain.

Place the tuna on top of the mixed greens.

Add the dressing, then the salt and pepper and finish with the cheese.

Chill before serving.

Roasted Chicken with Citrus and Garlic

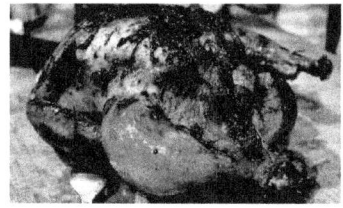

Ingredients:

1 whole free range, whole chicken

1 cup non iodized salt

1 orange

Some water to cover the bird in a stock pot

Roasting the Chicken

1 whole free range, whole chicken

1 stick pasture butter, sliced into 1 inch thick slices

1 lemon

1 large orange

3 sprigs rosemary, finely chopped

6 cloves garlic, finely chopped

Few drops of Extra Virgin Olive Oil

Instructions:

Brining the Chicken

Remove the giblets and put the chicken in a large stock pot and cover with water.

Boil a cup of water in a separate pan adding one cup of salt until all the salt is dissolved.

Add the boiled salted water together with the orange juice (include the rind) to the large stock pot with chicken and allow to marinate for at least3 hours but no more than 5 hours inside the refrigerator.

Roasting the Chicken

Preheat the oven to 425.

Lightly rinse the marinated chicken and pat dry.

With your fingers, make a 'butter pocket' by separating the chicken skin from the breasts.

Place the butter slices to the 'butter pockets' under the skin together with half of the chopped garlic.

Insert the oranges, lemons, the remaining garlic and rosemary inside the chicken cavity. Leave some rosemary for garnishing.

Brush the skin of chicken with a a little olive oil then sprinkle with the left over rosemary.

Roast the chicken at 425 degrees for an hour and a half.

After roasting the chicken pull it out of the oven and brush the outside skin with melted butter in the baking pan.

Carve and serve.

Paleo Chicken Salad Wraps

Ingredients:

1/2 cup chopped cooked or boiled chicken

3 tbsp chopped Fuji apples

2 tbsp chopped red grapes

2 tsp honey

2 tbsp almond butter

Instructions:

Make an easy Paleo chicken salad by mixing all ingredients together.

Wrap the chicken salad in a Romaine leaf and serve.

Paleo Beef Taco Salad

Ingredients:

1 lb lean ground beef

Rock salt

Black pepper

1 tbsp chili powder w/o preservative

2 to 3 garlic cloves crushed

3 large tomatoes (cleaned and stem removed) diced

1 head iceberg lettuce - rinsed, dried, and chopped

1 green bell pepper, diced

1 bunch green onions, chopped

1 yellow bell pepper , diced

12 ounces shredded almond cheese

1 to 2 avocado, sliced

Gluten free taco chips

Instructions:

Sear the beef in a skillet.

Drain the fat and dash with chili powder.

Set aside and allow to cool.

Combine romaine lettuce, iceberg lettuce, green onions, green pepper, jalapeno peppers, red or yellow pepper, tomatoes, and half of the cheese in a large bowl.

Toss well, put the cover on and refrigerate.

Put the salad on a bed of taco chips and top with the remaining almond cheese.

If taco chips that are gluten free are unavailable top with Paleo Mayonnaise instead and serve.

Serves 12

Beef Salsa Verde

Ingredients:

1 1/2 pounds sirloin steak

5 cups baby field greens

1 ripe soft avocado pitted

1 anchovy fillet, cut into pieces

1/2 C basil leaves

1 tomato sliced

2 cloves garlic

1/2 C parsley leaves

1 cup extra virgin olive oil

2 tbsp capers, drained

2 tbsp fresh lime juice

Freshly ground pepper

Sea salt

Instructions:

Make a puree out of parsley, tomato, garlic, basil, anchovy fillet, and capers using a food processor.

Add the olive oil slowly, pulsing the blender every now and then until all the ingredients are totally blended together.

Stir in the lime juice. Add enough pepper and salt as you wish and continue to blend until the sauce mixture has a smooth texture.

Taste and season as needed.

Set aside the sauce.

Add seasoning to the steak according to your taste and broil, or grill, or fry. Whatever way you like it done.

Cut the steak into thin slices and sprinkle with salsa verde sauce.

Top with avocado slices and serve over baby greens.

Paleo Nutty Meatloaf Recipe

Ingredients:

1 lb ground beef (you can use ground turkey or ground chicken as substitutes)

5 cloves garlic, minced

1 small green pepper finely chopped

1 tsp basil finely chopped

1 tsp rosemary finely chopped

1 tsp thyme finely chopped

½ cup of mixed ground almonds, walnuts, and pecans

2 eggs

black pepper, ground

Instructions:

Put all the ingredients in a big mixing bowl. Be sure the ingredients are evenly combined in the mixture.

Place the mixture in the refrigerator for 30 minutes to 1 hour.

Grease a loaf pan using olive oil and transfer the mixture into it.

Bake for one hour and 15 minutes.

Slice and serve.

Paleo Steak with Mushroom and Onion Gravy

Ingredients:

3 pieces 6-ounce Sirloin steak (tenderloin steaks are also good for this recipe)

1 cup coconut milk

2 tbsp coconut flour

1 medium onion sliced

2 to 3 garlic cloves crushed

1/8 tsp cayenne pepper

1/4 tsp sea salt

1/2 cup sliced mushrooms

1/2 tsp black pepper

1 tbsp coconut oil

Instructions:

Preparing the Steaks:

Broil, grill, or panfry the steak to your preferred doneness.

Preparing the Gravy:

Coat a skillet with cooking oil and heat at high setting.

Stir in onions, mushrooms, and garlic. When tender, remove from the skillet and put it aside for a awhile.

Add the coconut flour into the skillet and stir in the coconut milk. Continue stirring at high heat until the mixture is smooth.

Reduce the heat when the mixture is smooth and add in the seasoning.

Add the vegetable to the now almost thick gravy.

Simmer for another 5 to 10 minutes.

Coat the cooked steaks with the thick gravy generously and serve.

Beef and Spinach Salad

Ingredients:

1 pound spinach (chopped)

1/2 C red onions, sliced

1/2 pound beef, finely sliced

½ cup toasted cashews (You may use pine nuts, sesame seeds or cashew nuts instead)

Ground black pepper

1 teaspoon cayenne pepper

sea salt

1 tbsp coconut oil

Instructions:

Heat coconut oil in a saucepan over medium fire.

Put in the rest of the ingredients and cook/toast lightly. Stir well.

Take it off the fire and allow let cool.

Prepare the Balsamic Dressing by combining honey, balsamic vinegar and olive oil.

Sprinkle with the salad dressing and serve.

Paleo Pot Roast

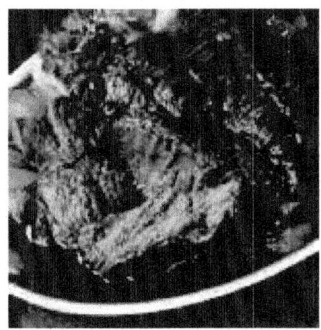

Ingredients:

3 pounds rump roast, with fat trimmed

2 tbsp coconut oil

1 onion, large, chopped

2 celery ribs, chopped

1/2 tbsp dried thyme

3 cloves garlic, minced

2 C beef broth (without preservative)

1/2 tsp dried parsley

1 to 2 bay leaves

20 whole peppercorns

1/2 tbsp sea salt

6 to 7 florets of cauliflower

2 carrots, sliced

1/3 C coconut flour

1/2 tsp sea salt

Instructions:

Heat coconut oil on a Dutch oven or any thick walled cooking pot.

Add the rump roast and sear all sides.

Remove the roast from the Dutch oven and put in a platter.

 Put the celery, parsley, onion, thyme and garlic into the Dutch oven and sauté for 5 minutes;

Put the seared rump roast back to the Dutch oven.

Add the broth, bay leaf, sea salt and peppercorns. Cover and place inside the oven preheated to 3250 F.

Cook the roast for 4 hours basting it every half hour.

Remove roast after it is done and strain the stock into a bowl.

Throw away the vegetables keeping only the stock.

Shred the meat with the use of two forks and put the shredded meat back into the Dutch oven.

Immerse the shredded beef with the previously strained liquid from the pot.

Throw in the cauliflower, carrots and the remaining sea salt.

Put inside the oven again and simmer for another 45 minutes.

Drain the stock from the Dutch oven and measure. Add more beef broth if necessary to make 3 cups of stock then pour all 3 cups into the saucepan.

Stir in the flour and simmer

until it turns into thick gravy.

Cover the meat and vegetables with the gravy.

Paleo Grilled Steak Portobello Gravy

Ingredients:

1 Sirloin or tenderloin steak

2 tbsp Coconut Flour

½ tsp ground black pepper

1 tsp seasoning salt preservative free

1 tsp sea salt

2 tbsp olive oil

3 to 4 cups hot water (as needed)

1 white or red onion sliced

1 cup Portobello mushrooms sliced with stem

1 cup chives slices

2 to 3 garlic cloves (finely chopped)

Instructions:

Preparing the Gravy

Coat a skillet with olive oil and bring to high heat.

Stir in coconut flour gradually adding water while stirring.

Continue to stir to remove lumps and until the mixture is smooth.

Reduce the heat as the gravy starts to thicken.

Add pepper, seasoning salt and sea salt to taste.

Stir in mushrooms, chives, garlic and onions.

Simmer at low heat until vegetables are tender.

Preparing the Steak

Rub the steak with black pepper, seasoning salt and sea salt.

Coat a skillet with olive oil and bring to high heat.

Grill the steak on the skillet until both sides are cooked to desired doneness. (rare, medium or well done)

Place the steak on a serving plate and cover with Portobello gravy. Serve hot.

Paleo Steak Kebobs

Ingredients:

1 lb lean steak (1 inch cubes)

2/3 C Olive Oil

3 tbsp beef spice rub (preservative free)

1/3 C lime or lemon juice

2 medium size zucchini (cut into 1 inch rounds)

2 cloves garlic, finely chopped

1 large red or white onion (wedged)

2 red or green peppers (cored and seeded, cut into 1 inch pieces)

Sea Salt and Ground Black Pepper To Taste (optional)

8 to 12 Metal or Bamboo Skewers.

Instructions:

For the marinade, combine lemon juice, olive oil, garlic, and spices in a large bowl. Reserve one fourth cup for use in basting when grilling.

Combine all the other ingredients including the beef cutlets in the bowl of marinade and toss lightly enough to coat everything evenly.

Cover the bowl and place inside the refrigerator.

Allow the steak to marinate for up to six hours.

Place the marinated beef, onions and bell peppers on skewers. Set aside the marinade for use later as basting.

Arrange the kebobs on the char-grill that has been preheated earlier. Don't forget to oil the grill before placing the kebobs.

Turn the kebobs over every two minutes while basting the

done sides for the glaze.

Cook to desired finish (rare, medium rare, well done). Add pepper and pepper before taking them off the grill.

Garnish with lime or lemon halves and serve.

Beef Roll

Ingredients:

1 ½ pound ground beef

1 egg

½ red or white onion (peeled and diced)

Black pepper to taste

Sea salt to taste.

½ tbsp olive oil

1/8 cup finely diced parsley

Instructions:

Beat egg well in a small bowl and set aside.

In a separate bowl mix the ground beef with diced onion, pepper and salt.

Add the egg into the meat mixture and mix well.

Roll portions of the meat mixture into medium sausage sizes.

Line sauce pan with olive oil and heat at medium-high.

Add each beef roll and grill until thoroughly brown all over.

Remove the beef rolls from the pan and place on a plate covered with paper towel. Let it stand until the excess fat drain completely.

Arrange the beef rolls on top of your favorite vegetable.

Garnish with fresh parsley flakes and serve.

Shrimp, Cantaloupe & Mint Salad

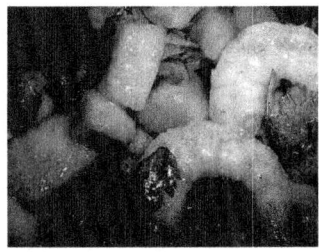

Ingredients:

3 C fresh Arugula

1 C ripe mango cubes, fresh

1 1/2 lbs. medium shrimps, pre-cooked

3 tbsp fresh lemon juice (you can use lime)

1 C cantaloupe cubes, fresh

1 tsp nutmeg

1 tsp cinnamon

Instructions:

Mix shrimp, cantaloupe, mango, nutmeg, cinnamon, and lemon juice.

Top the salad mixture with your choice of balsamic dressing.

To prepare the Balsamic Dressing:

Combine olive oil, honey and balsamic vinegar in a bowl.

Serves 4

Paleo Prawns Tomato Sauce

Ingredients:

2 pounds prawns deveined unpeeled

3 cloves garlic minced

2 tablespoons olive oil

4 medium fresh tomatoes, stem removed and finely chopped

1/4 cup fresh parsley

1 tsp sea salt to taste

1 tsp black pepper to taste

½ tsp cayenne pepper to taste

½ cup white or red onion finely chopped

1 tsp oregano

2-3 stalks celery finely chopped

1-2 tbsp capers

Instructions:

Coat a skillet with olive oil and heat to medium-high.

Simmer garlic, parsley, onion, and celery until tender.

Add in chopped tomatoes and seasoning to taste and continue to simmer at low heat for 15 to 20 minutes.

Add the shrimps and continue to cook for another 3 to 5 minutes or until the prawns are no longer pink and translucent. Be careful not to overcook shrimps.

Garnish with capers and serve.

Paleo Grilled Salmon with Asparagus

Ingredients:

1 lb salmon

1/2 lb cherry tomatoes

1/2 lb asparagus

1 tsp sea salt

1 tbsp olive oil

1 tsp black pepper

Ingredients:

Line a skillet with olive oil and sauté until tender the cut asparagus.

Add pepper and salt to taste. Set aside while you prepare the salmon.

Clean fresh salmon and season according to your taste.

Broil salmon making sure it is thoroughly cooked from top to bottom. Once cooked remove salmonfrom skillet and set aside.

Place the sautéed asparagus on a serving plate and place the cooked salmon on top of it.

Garnish with ½ lb cherry tomatoes and serve.

White Fish w/ Macadamia Salsa

Ingredients:

fillets of 2 whole white fish

½ cup chopped tomatoes

¼ cup macadamias, halved

1 avocado, diced

3tbsp parsley, chopped

3 tbsp coriander, chopped

Olive oil

Instructions:

Set grill temperature to medium and pre heat.

Grill fish for 3 to 4 minutes till cooked.

For the salsa, combine macadamias, coriander tomatoes, parsley, and avocado in a bowl. Toss.

Add olive oil while tossing making sure everything is well coated.

Place the grilled fish on a serving plate.

Add the salsa and serve.

Paleo Pineapple Glazed Salmon

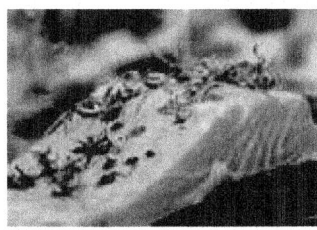

Ingredients:

Glaze the Ingredients

1 eight ounce can of Crushed Pineapple in Juice

Half Cup Sweet Chili Sauce

1 Tsp Cajun Seasoning (w/o preservative)

8 to 12 pineapple slices (can be canned or fresh)

Fish Ingredients:

4 pieces of 5 to 6 ounce Salmon Fillets

Pepper and Salt, to taste

1 tbsp coconut oil

Instructions:

Clean and season the salmon with rock salt and pepper.

Place the salmon on a non-stick 9 X 13 baking dish. If you don't have a non stick dish use any ordinary dish of the same size but with nonstick cooking spray.

Place the dish inside an oven that has been preheated to 400 degrees and bake the salmon for 15 to 20 minutes or until the salmon flakes easily when forked.

Combine all the Glaze ingredients in a small bowl and mix well.

Cover each piece of Salmon with a generous scoop of the Pineapple Glaze.

Top the cooked salmon with fresh pineapple slices and serve.

Serves 4

Paleo Chicken Stir Fry

Ingredients:

1 lb boneless chicken cutlets

1/2 lb broccoli florets

5 cloves garlic finely chopped

1/4 lb red bell peppers (sliced)

2 tbsp coconut oil

1/4 lb fresh carrots (diameter cut)

1/4 lb finely chopped chives

sea salt and black pepper to taste

Instructions:

Put a little coconut oil to a frying and sauté chicken cutlets until browned and cooked thoroughly.

Put the cooked chicken cutlets aside.

Heat coconut oil in another frying pan.

Add the broccoli, carrots, red pepper, garlic and chives.

Simmer until the vegetables are tender. Add the chicken cutlets and season with pepper and salt to taste.

Place in a serving plate and serve.

Paleo Jalapeno Glazed Chicken

4 skinless and boneless chicken breast

4 jalapenos (chopped)

1 medium pitted and sliced green bell pepper

1 medium pitted and sliced yellow bell pepper

2 teaspoons fresh chopped parsley

2 teaspoons chili powder

1 teaspoon thyme

1 teaspoon ground cumin

2 teaspoon sea salt to taste

1 teaspoon black pepper

2 tablespoons raw honey

Juice 2 fresh limes

1 cup chicken broth

Instructions:

Combine chili powder, thyme, cumin, sea salt and black pepper in a small bowl. Mix well to make the spice mixture.

Rub each piece of chicken breast thoroughly with the spice mixture.

Place in a baking sheet, cover, and place inside the refrigerator for two hours.

In a sauce pan, combine chicken broth, lime juice, honey, sliced jalapenos, green and yellow bell peppers and parsley and cook over medium-high heat until the vegetables are tender and the mixture thickens.

Arrange the chicken cutlets in a baking sheet without the marinade.

Place the baking sheet with the chicken cutlets inside an oven that was preheated to 450 degrees.

Cook for 15 minutes and baste the chicken with the marinade juice.

Cooked for another 45 minutes or until the chicken is thoroughly cooked.

Serve with Cauliflowers instead of rice.

Here is the Paleo Cauliflower Recipe:

Ingredients

1 medium cauliflower

Sea salt to taste (optional)

Keep the cauliflower refrigerated and unwashed until ready to cook.

When ready to cook, wash and remove the outer leaves.

Cut a deep X pattern n the core of the cauliflower.

Place the whole cauliflower stem first in 3 1/2 quarts of salted boiling water.

Continue to boil 20 to 25 minutes.

Remove the cauliflower let it cool.

Once cooled, chop the cauliflower to a rice-like consistency.

Serve

Baked Chicken with Pomegranate Glaze

Ingredients:

1 whole chicken (about 4 to 5 lbs)

1 large lemon (pierced with fork)

1 tbsp Dijon mustard

2 sprigs rosemary, fresh

Seeds from 1 pomegranate

1 tbsp finely chopped garlic

2 sprigs fresh thyme

2 cups pomegranate juice, unsweetened

2 tbsp plus 1 tsp arrowroot

1 tsp honey

1/2 sea salt 1/2 tsp freshly ground black pepper

Instructions:

Place pierced lemon and rosemary inside the chicken cavity.

Tie the chicken legs together to keep it firm and set in a roasting pan.

Mix together pomegranate juice, mustard, thyme, arrowroot, and garlic for basting.

Pour the mixture all over the chicken and sprinkle with salt and pepper.

Place inside an oven preheated to 375 degrees.

Bake for 25 minutes. Take it out and baste the chicken.

Bake for another 25 minutes and then baste again.

Add the pomegranate seeds and reduce oven temperature to 350 degrees Fahrenheit.

Bake for another hour while basting every half hour this time.

When done drain the liquid and set aside.

Cover the chicken with foil and let stand for 30 minutes.

Serve the chicken with the glaze.

Paleo Roasted Chicken and Herbal Gravy

Ingredients:

5 to 6 boneless chicken breast (with skin)

1/2 tsp dried thyme

1 quart organic low sodium chicken broth

2 tbsp fresh rosemary (sprigs)

4 to 5 tbsp olive oil or coconut oil

8 crushed garlic cloves

2 to 3 tbsp coconut flour or almond flour.

sea salt and pepper to taste

Instructions:

Cover chicken breasts with olive oil and season with pepper and salt.

Arrange in a baking sheet that has been lined with aluminum foil.

Place inside the oven preheated to 350 degrees.

Bake without cover until the chicken is completely cooked (about 30 minutes).

Take it out of the oven and cut the chicken breasts horizontally into 2 inch slices.

For the Gravy

Heat coconut oil in a sauce pan over high heat.

Pour in the chicken broth, thyme, rosemary, salt, pepper and garlic.

Mix well and gradually reduce heat to low.

Continue to simmer until you have the desired texture and consistency.

Place roast chicken breast, mix well, and then remove from fire. Serve hot.

Paleo Pumpkin and Chicken Curry

Ingredients:

2 pieces sliced chicken breasts

2 tablespoon olive oil

5 cups pumpkin, diced

2 garlic cloves, finely chopped

1 bunch fresh coriander, chopped

1 onion, diced

2 tbsp ginger, ground

1 tbsp turmeric, ground

2 tbsp coriander, ground

2 tbsp cumin, ground

1 ½ cups vegetable stock

Salt

Instructions:

Sauté garlic and onion in a Fry pan for two minutes over medium heat.

Add chicken and simmer while stirring frequently for ten minutes until chicken turns white.

Stir in the pumpkin, cumin, turmeric, ginger, and coriander and continue to stir for one minute.

Pour the vegetable stock and simmer for 15minutes on low heat.

Stir in the chopped coriander then put the cover on and simmer for another two minutes.

Add salt to taste.

Paleo Crispy Orange Chicken

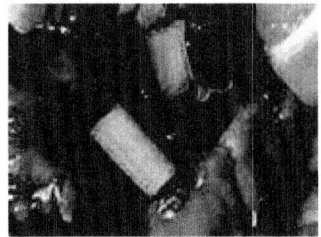

Ingredients:

2 pounds chicken (deboned, skinless)

1 1/4 tsp pepper

1.5 tsp sea salt

.5 C coconut meal/flour

1 tbsp coconut oil, extra virgin

1 egg

For the Glaze:

1 teaspoon garlic, minced

1 cup fresh orange juice

1.5 tsp grated orange rind

.5 C hoisin sauce

1/4 cup honey

Dash of cayenne pepper

Sea Salt

Pepper

Instructions:

Cut the chicken into two inch pieces and place them in a large container or a mixing bowl.

Add the egg, coconut oil, pepper and sea salt. Combine well and put aside.

Put the ½ cup coconut meal/flour mix in a separate bowl and dip the chicken on the flour mixture making sure each piece is coated generously.

Spoon enough coconut oil to fill a large skillet to about half an inch from its bottom. Put the skillet on the stove and set the temperature on high heat until it reaches 3750 F. Start frying the chicken pieces by batches. Fry each batch until the chicken cutlets are browned, crispy, and crunchy.

This should be about 3-4 minutes per batch.

Take out the chicken pieces from the skillet and drain them well over paper towels.

Finish frying the rest of the chicken pieces and set them aside while you prepare the glaze.

Remove all the oil from the skillet leaving only about 2 tbsp. of oil and lower the temperature to medium heat.

Sauté the garlic for one minute but avoid burning it otherwise it will give a bitter taste.

Toss in all the other ingredients and let the mixture boil.

Stir the boiling mixture for three minutes before reducing the heat. Continue to simmer the mixture until you produce the glaze.

Cover the chicken with the glaze and decorate with chives and orange slices before serving.

Chicken and Egg Salad with Almond Satay Sauce

Ingredients:

2 hard boiled eggs,(quartered)

1 steamed chicken breast, shredded

1 carrot, diced

2 cups rocket leaves

½ green capsicum, diced

10 cherry tomatoes (halved)

For the Almond sauce

1 small white onion, grated

1 tablespoon oil

2 crushed garlic cloves,

¼ cup coconut cream

¼ cup almond butter

Chili flakes

1 tablespoon hoisin sauce

Instructions:

Combine shredded chicken, rocket leaves, diced carrot and capsicum in a mixing bowl and mix well. Set aside

Coat a frying pan with oil and set on low heat.

Sauté onion and garlic until lightly browned (about 5 minutes).

Add in the almond butter, hoisin sauce, and coconut cream stirring continuously until the mixture thickens.

Remove the pan from the heat and mix in the chili flakes.

Allow to cool a bit before removing the excess oil.

Set the rocket salad over a serving plate and top with egg slices.

Top the salad with the almond sauce and serve.

Paleo Dinner Recipes

Paleo Duck Breast with Cherry Chutney

Ingredients:

1 tbsp olive oil (extra virgin)

½ C small onion (chopped)

3 cloves crushed garlic,

½ tsp black pepper

1 tbsp shallot, finely chopped

½ tsp ground cumin

¼ tsp red pepper flakes, dried

¾ tsp sea salt

1 coarsely chopped

plum tomato

½C red bell pepper coarsely chopped)

¼ C red wine, dry

2 tbsp raw honey

1 ½ to 2 tbsp apple cider vinegar

½ tsp mustard (no soy)

3 C Bing cherries(1 can)

½ C Golden Raisins

6 oz boneless duck breasts with skin (about 10 pcs.)

2 tbsp water

1 tbsp tarragon or chives (chopped, fresh)

Instructions:

To prepare the glaze and the chutney

Cover a large heavy saucepan with oil and set over moderate heat till it is hot enough but short of smoking.

Sauté the shallot, onion and garlic together until golden (approximately 7 minutes). Do not forget to stir occasionally.

Add in the tomato paste, cumin, black pepper, hot pepper flakes and 1/4 tsp salt and cook for 30 seconds while stirring continuously.

Reduce the heat to medium before stirring in the bell pepper. Continue to cook for five minutes this time.

Pour in the wine, and enough vinegar to suit your taste together with the raw honey.

Allow to boil lightly for 5 minutes more then add the mustard, the 1/2 teaspoon salt and the cherries.

Simmer for one minute more.

Put aside to cool a bit.

Take ¼ cup of the now slightly cool mixture and pour into a blender. Pulse for one minute to make a puree that is extra smooth. Set aside to be used for glazing duck later on.

To complete the chutney, add the remaining one and a half cups of cherries, chives, tarragon, and the golden raisins.

Ready the duck for cooking by scoring the duck skin with a sharp knife making a crisscrossing pattern all over.

Rub the duck thoroughly with pepper and salt.

Pour water into a heavy 12-inch skillet and set over low heat until hot.

Put the duck in the skillet with the skin side at the bottom.

Cook uncovered on low heat for about 25 minutes or until almost all of fat has been melted (rendered) and skin has turned golden brown. Do not turn or move the duck while cooking.

Remove the duck from the skillet and transfer to a serving plate.

Take out most of the fat from skillet leaving only about 1 tablespoon of fat on it.

Brush cherry glaze all over the duck and put back into the skillet with the skin side on top this time.

Put the duck back into the oven and roast for 8 minutes or until the oven thermometer reaches 135 degrees Fahrenheit. The duck should be done medium-rare by then.

Remove from oven and let it rest for five minutes.

Cut the duck into slices with the knife at a 45-degree angle. Use a very sharp knife for clean cuts.

Decorate with vegetables and fresh fruits and serve with the cherry chutney.

Grilled Prawns

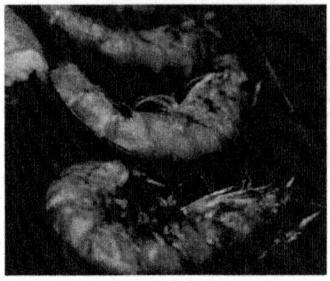

Ingredients:

1 lb of prawns (you can use large shrimps instead)

Juice from 4 fresh lemons or limes

4 cloves garlic, crushed

1 small red onion, minced

2 tsp olive oil

Sea salt

Black pepper

Instructions:

Devein the prawns but do not remove the shell and the tail.

Wash clean the prawns and allow to dry.

Prepare the marinade by combining the lime juice, minced onion, crushed garlic, olive oil, pepper and sea salt. Mix everything well.

Soak the deveined prawns in the prepared marinade and marinate for at least 3 to 4 hours.

Cook/Grill the two sides of prawns with olive oil for about 5 minutes.

The prawns cook quickly so watch it carefully to avoid overcooking.

Spinach Salad with Aragula, Peaches,Berries & Nuts Salad

Ingredients:

1 1/2 C fresh Arugula

3/4 cup slice peaches

1 1/2 cups spinach

1/2 cup blackberries

1 tsp cinnamon

1/2 cup walnuts soaked

4 to 6 slices bacon (fried)

2 tablespoons coconut oil (extra virgin)

1 Boiled egg

Instructions:

Wash the arugula, spinach and blackberries thoroughly then set aside to dry.

Halve walnuts with a sharp knife.

Sauté the slices of fresh peaches using extra virgin coconut oil till tender. Put the peaches aside.

Crumble the crisp fried bacon into bits.

Shell the boiled eggs and chop finely.

Toss the arugula, spinach, and blackberries with the bacon bits, egg bits, walnut halves, nutmeg and cinnamon together.

Top the salad with the sautéed peaches and drizzle with the balsamic dressing of your choice.

Egg and Capsicum Salad

Ingredients:

2 eggs, hard boiled, diced

1tbsp coconut oil

2 diced bacon eyes

½ diced green capsicum

¼ C chopped parsley

1tbsp mayonnaise

1 C salad leaves, mixed

Instructions:

Cook bacon in a coconut oil coated frying pan over medium heat. Fry the bacon until it starts to be crispy.

Drain the excess oil and transfer the bacon to a bowl.

Add the boiled eggs, parsley, capsicum, and mayonnaise. Toss well.

Arrange the salad on a serving plate and top with egg and capsicum mixture to serve.

Sautéed Chard, Golden Raisin and Pine Nut

3 C chard's (sliced)

1 1/2 tbsp olive oil

1 tsp black pepper

1/2 cup pine nuts

1/2 tsp sea salt

3 cloves garlic(finely chopped)

1/2 C golden raisins

Instructions:

Season the slices of chard's with pepper and salt.

Sauté together with the garlic until tender.

Take away from the heat when the chard's slices are tender.

Mix the chards with pine nuts and golden raisins.

Spinach, Apple, Glazed Walnut Salad

Ingredients:

1/4 C honey (raw)

white of 1 egg

1 tsp cinnamon

2 C walnuts (halved)

1 tsp nutmeg

Slices of 1 cored apple

1 tbsp lime or lemon juice

1/2 C leaves of baby spinach

Instructions:

Preparing the Glaze

Heat raw honey in a small pan over low heat.

Combine honey with nutmeg and cinnamon in another bowl. Put it aside for a while.

Whisk egg whites till it foams with stiff peaks.

Coat the walnut halves with foamed egg whites. Pour in the honey, nutmeg, and cinnamon mixture. Make sure the walnuts are covered evenly.

Arrange the walnuts evenly over a baking sheet that has been greased earlier. Bake for one hour in an oven that was preheated to 250 degrees.

Stir the walnuts after the first 30 minutes.

Preparing the Apple

Wash the apple and take out the core before slicing into the desired size.

Sprinkle the slices of apple with lime or lemon or and set aside.

Preparing the Salad

Line 2 salad dishes with the leaves of baby spinach.

Put the fresh apple slices on top of the spinach.

Spread the glazed walnuts over the salad.

You may add seeds of your choice and serve.

Paleo Asparagus Mushroom

Ingredients:

2 C asparagus (cut in quarters)

1 C mushrooms (trimmed and quartered)

1 tbsp olive oil

2 sprigs of rosemary (fresh, minced)

Sea salt

Black pepper

Cayenne pepper

Instructions:

Drizzle olive oil over a frying pan and place over high heat.

Sauté mushrooms and asparagus together adding enough seasonings to suit your taste.

Take out from the fire when vegetables are done.

Serve hot.

Paleo Marinated Mushroom

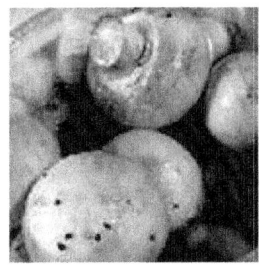

Ingredients:

1 lb mushrooms (small)

1/3 C red or white wine vinegar

1/3 C olive oil (extra virgin)

1 red onion (chopped fine)

1 tsp sea salt

Dash of cayenne pepper

1 tsp black pepper, ground

1 tsp oregano fresh or dried, minced

1 tbsp parsley, fresh or dried

1 tbsp rosemary, fresh or dried

3 cloves garlic, peeled, finely chopped

raw honey

Instructions:

Wash and clean mushrooms thoroughly and cut off the stems. Put aside.

Heat olive oil in a sauce pan and leave simmering for a while before reducing the heat.

Add in all the ingredients (except the mushrooms and bring to a boil.

Reduce heat to low and stir in the mushrooms.

Simmer for about 5 to 10 minutes until the mushrooms are tender.

Take off from the heat and place in a covered container.

Put inside the refrigerator for at least an hour. Serve it chilled. Serves 12.

Asian-Style Beef Short Ribs

Ingredients:

5 pounds beef short ribs

1 tsp fat/oil

5 cloves of garlic

2 tbsp ginger root, fresh, peeled

1 onion, chopped

1 Serrano Chile pepper

1/2 C coconut aminos

2 C beef stock

1/4 C rice wine vinegar (gluten-free)

2 C water

2 tbsp fish sauce

5 oz shiitake mushrooms, fresh

1 juiced orange, zested

3 to 4 thinly sliced green shallots

salt

pepper

Instructions:

Rub the beef ribs lightly with pepper and salt.

Coat an oven proof large pan thinly with 1 tsp fat/oil so the ribs will not stick to the pan. Set the pan over medium-high fire.

Broil the ribs for no more than 2 minutes on each side. Broil the ribs by batches. Place the seared ribs in a platter and keep warm by covering them with foil.

In the same pan, cook the chopped onion for ten minutes.

Meanwhile mince or chop the garlic, pepper and ginger and add to the pan together with the sautéed onions. Cook for one minute while stirring continuously.

Put the seared beef ribs back into the pot and pour the beef stock, coconut aminos, water, gluten free vinegar and fish sauce. Make sure the liquid covers the ribs well.

Cook without the lid and bring to a simmer before taking the pot out of the fire. Cover the pot tightly this time and place inside an oven that has been preheated to 300 degrees F. Cook for 3 hours.

Remove the pot from the oven after 3 hours. Take out the short ribs and place on a clean platter. Keep the ribs warm by covering them with foil again.

Allow the liquid inside the pot to cool off a little before straining it with a sieve. With a spoon, press the solids to squeeze out the juices. Discard the solids and skim off most of the fat from the strained sauce.

Put the sauce back to the pot and simmer until you reduce volume to about one and a half cups. Add pepper and salt to taste.

Add the orange juice and the shiitake mushrooms and continue simmering for about 5 to 10 minutes until the mushrooms are tender.

Place the beef short ribs back into the pot and coat them with the sauce.

Garnish with sliced scallions and orange zest before serving. Serve the ribs together with the mushrooms.

Crispy Duck Breast with Chipotle Orange Sauce

Ingredients:

For the Duck

4 half pound duck breasts (halved and with skin)(Peking Duck from Long Island is best)

Kosher salt

black pepper, fresh ground

For the Chipotle Sauce

3 C fresh orange juice

1 tbsp maple syrup

1/3 C lime juice, fresh

1 tbsp chipotle pepper with adobo sauce, finely chopped

1 cinnamon stick

3 cloves of garlic

1 tsp kosher salt

2 tbsp duck fat, melted

Instructions:

Preparing the Chipotle Sauce

Combine all the listed ingredients for the Chipotle Sauce in a heavy duty saucepan and place over medium fire.

Allow it to simmer for 20 to 30 minutes until you only have about 1 cup of sauce left. Cool the sauce and strain using a sieve.

Preparing the Duck Breast

Wash and dry the duck breasts. Pat dry with paper towels.

Use a sharp knife to score or cut a crisscross pattern across the skin and of the duck without cutting through the flesh. Sprinkle with black pepper and Kosher salt.

Set a pan over a stove on medium to high heat and place the duck breast, with the skin side down, to sear it. You need not coat the pan with oil before placing the duck breast on it as the duck fat will start to render because of the heat.

Lower the heat to medium and continue to broil the duck breasts until the skin has turned brown and crispy by which time most of the fat coming from the duck breasts has melted or rendered into the pan. This should be within 4 minutes. Turn over the breasts and continue cooking until you get the desired doneness.

When done, take the duck out of the pan and and place on a chopping board or plate to rest and cool for five minutes before slicing.

Take 2 tablespoons of the melted duck fat and add them to the Chipotle Orange Sauce. Serve the duck slices together with the sauce while it is still warm.

Serves 4-6

Dry Rubbed Barbecue Pork Ribs

Ingredients:

For the Dry Rub

1/4 C coconut crystals

2 tbsp paprika, smoked

2 tbsp garlic powder

1 tbsp mustard, dry

2 tbsp onion powder

1 tsp black pepper

2 tbsp sea salt

2 tsp cayenne pepper

1 tbsp dry basil (crushed)

For the Ribs

3 racks of full baby back ribs

For the Barbecue Sauce

1 1/2 C of the juices used in cooking the ribs (or you use beef stock instead)

3 cloves fresh garlic, finely diced

1 can tomato paste (6 oz)

1/4 C apple cider vinegar

2 tbsp honey

1/4 C mustard

1 tsp onion powder

1/2 tsp ground cumin

1 tsp chili powder

black pepper to taste

1/2 tsp sea salt, coarse

1/2 tsp paprika, smoked

1/2 tsp cayenne pepper

Instructions:

Preparing the ribs

Wash and drain the ribs. Dry them with paper towels.

Arrange the ribs on baking pans with shallow sides. Generously coat both sides of the ribs with the dry rub making sure they are coated evenly all over.

Wrap the ribs in aluminum foil folding and pinching all edges for a tight seal to retain the juices inside.

Allow it to marinate before cooking for 30 minutes.

Broil the ribs for four hours at 2000 Fahrenheit.

After four hours turn the oven off but leave the ribs to "rest" inside for at least 30 minutes.

Meanwhile you can heat up the grill in the last 10 minutes. Then take the foil off the ribs slowly while pouring the juices inside each wrap into a container. You will use this later as barbecue sauce.

Arrange the ribs on the grill and set the temperature between medium and low.

Close the lid and grill for 10 minutes.

Turn over the ribs and grill another 10 minutes.

Take the ribs out of the grill and wrap again with foil. Allow to rest for fifteen minutes before serving.

Preparing the barbecue sauce

In the meantime the ribs are on the grill you can start preparing the sauce. Combine the collected drippings with diced garlic in a sauce pan (medium size) and simmer for five minutes. Add in all the other ingredients and boil lightly on medium to low heat for twenty minutes or until the sauce is thick enough for you.

Paleo Bread Recipes

Paleo Bread

Ingredients:

1 1/2 C almond flour, blanched

2 tbsp coconut flour

1/4 cup golden flaxseed meal

1/4 tsp sea salt

1 1/2 tsp baking soda

5 eggs

1/4 cup coconut oil

1 tbsp honey

1 tbsp apple cider vinegar

Instructions:

Place the almond flour, baking soda, coconut flour, flax, and salt in a food processor and blend the ingredients well.

Pulse in the eggs, oil, honey and vinegar in that order. Continue to pulse until you produce a smooth batter.

Pour the batter into a greased non-stick 7.5" x 3.5" loaf pan. If you don't have a non stick loaf pan you can line and ordinary loaf pan with parchment paper. This will do the trick.

Bake for about 40 minutes at 350° F.

Cool and serve.

Date and Walnut Bread

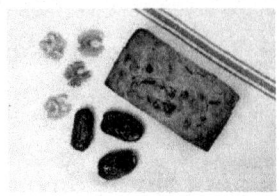

Ingredients:

½ C almond flour, blanched

2 tbsp coconut flour

¼ tsp baking soda

⅛ tsp celtic sea salt

3 large Medjool dates, pitted

1 tbsp apple cider vinegar

3 large eggs

½ cup walnuts, chopped

Instructions:

With a food processor, mix almond flour and coconut flour and blend well together.

Add in baking soda and salt. Pulse to blend.

Add in dates and pulse again until the mixture has the same texture as coarse sand.

Stir in apple cider vinegar and eggs and blend well.

Pulse in the walnuts briefly.

Pour the batter on to a miniature loaf pan.

Put inside the oven and bake for 28 to 32 minutes at 350° F.

Cool the bread while still in pan for 2 hours before removing.

Makes 1 small loaf of Walnut Bread.

Dark Rye Bread

Ingredients:

1 cup almond flour, blanched

½ tsp baking soda

¾ tsp cream of tartar

3 eggs

2 tbsp olive oil

¼ cup water

1 tsp agave nectar or honey

1-2 tbsp caraway seeds

Instructions:

Mix almond flour, flax, salt, baking soda and cream of tartar in a large bowl.

In a separate, smaller bowl mix eggs, oil, water and agave.

Stir in the dry ingredients into wet. Add the caraway seeds and mix well.

Let the batter to stand for 1 to 2 minutes for it to thicken.

Pour the batter to a small, greased loaf pan (6.5 x 4 inch).

Bake for 30-35 minutes at 350° F.

Paleo Banana Bread

Ingredients:

1 ½ cups bananas (mashed)

1 tbsp honey

1 tbsp vanilla extract

3 eggs

¼ C vegan shortening

½ tsp sea salt

2 C almond flour, blanched

1 tsp baking soda

Instructions:

Combine bananas, honey, vanilla, eggs, and shortening into the food processor and pulse them together.

Add in the almond flour, baking soda, and salt while pulsing with each addition.

Pour the batter into a previously greased loaf pan (7.5 x 3.5) .

Bake for 55 to 65 minutes at 350°F.

Take out of the oven and cool before removing from the pan.

Cranberry Loaf Bread

Ingredients:

¾ C roasted almond butter , at room temperature

3 eggs (large)

2 tbsp olive oil

¼ C arrowroot powder

¼ tsp baking soda

almond flour(blanched) for dusting

½ cup cranberries (dried)

olive oil

¼ C dried apricots(chopped)

¼ C sesame seeds

1 tsp sea salt

¼ C sunflower seeds

¼ C pumpkin seeds

¼ cup plus 2 tbsp sliced almonds

Instructions:

Combine the eggs, olive oil and almond butter in a large bowl and blend together until smooth using a hand blender

In a separate medium size bowl, blend arrowroot powder, baking soda and salt.

Blend the wet mixture with arrowroot mixture until well mixed.

Fold in the sliced almonds, dried apricots, sesame seeds and cranberries.

Grease a loaf pan (7.5 x 3.5) with olive oil and powder with almond flour. Transfer the batter carefully into the pan and top evenly with whatever sliced almonds left.

Bake for 40 to 50 minutes at 350° or until a knife inserted into center comes out clean.

Cool the bread for at least 1 hour while in pan before removing.

Paleo Soup Recipes

Sopa de Lima

Ingredients:

4 halves chicken breast

¼ tsp Chili Powder

4 Cups of Organic Chicken Broth (you can use your own homemade broth)

¼ tsp Garlic Powder.

2 pieces Chile Peppers, Serrano

2 Tomatoes, chopped

4 to 5 Garlic Cloves, peeled and minced.

1 Tbsp Olive Oil

½ of an Onion, chopped

⅓ C Lime juice (or you can take the juice from 2 limes)

½ tsp Lime Zest

2 Tbsp Cilantro, chopped, fresh

1 peeled Avocado, chopped.

Sea Salt

Instructions:

Arrange chicken in a baking dish lined with oil. Sprinkle the chicken with garlic and chili powder.

Place the baking dish inside the oven preheated to 400 degrees and bake the Chicken for twenty minutes.

In the meantime the chicken is baking in the oven, set a stock pot over the stove and add the olive oil.

Sauté the Garlic, Onion and Serrano peppers together until they are tender for three minutes.

Stir in the Chopped Tomatoes and leave simmering for at least two minutes more.

Add lime juice and chicken broth while stirring occasionally. Adjust temperature to low heat.

Remove the chicken from oven after two minutes and cool just enough for you to be able to handle it and chop it into small bite sizes.

Return the chopped chicken pieces to the stock pot, and increase the heat to bring the mixture to boil. Once boiling, lower the heat to the lowest setting, and put on the lid cover.

Let it boil lightly for an additional 20 minutes on low heat.

Place the avocado chunks into each serving bowl (not directly into soup stock).

Put the avocado in each bowl first then pour the soup over it.

Finish by garnishing with Cilantro.

Bacon Soup with Asparagus, Green Peas

Ingredients:

1 bunch asparagus, medium, trimmed

1/2 white onion, chopped

4 slices of bacon

3 C chicken stock (you can use vegetable stock too)

2 cloves garlic, minced

1 C fresh peas (you can use frozen peas but defrost them first)

1/2 cup milk

Instructions:

(Note: green peas are sometimes allowed in the Paleo Diet as long as they are not consumed often and the in between periods between consumptions are long.)

Use a wide-bottomed frying pan to fry the bacon till they start to be a little crispy. Remove them immediately from the pan using a slotted spoon and put them aside to cool for a while.

Crumble or Chop the bacon into tiny pieces when cool enough.

Sauté the onion and garlic using the same pan till the onion turns golden. Add the peas and asparagus after.

Stream in the chicken stock and let it boil again. Cover and lower the heat. Allow to simmer for twenty minutes before taking the pan off the heat. Use a hand held blender to process the soup until it is smooth. Add seasoning according to taste. Top with bacon bits before serving.

Paleo Thai Seafood Stew with Sweet Potato, Basil and Lime

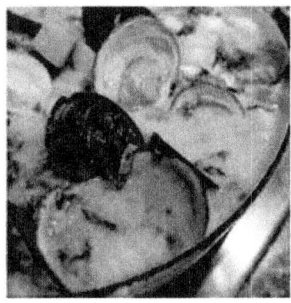

Ingredients:

2 tbsp Olive Oil

2 cups chopped Shallots

4 Cloves of minced garlic,

2 pieces Red peppers, chopped

3 tbsp Thai red curry paste

1/2 jalapeno pepper, minced, w/o seeds

28 oz coconut milk (2 cans)

1 C clam juice

1 tbsp honey

2 tsp fish sauce

1/2 tsp salt

2 Pounds wild caught cod, cut into 3" pieces

3 Large sweet potatoes, cut into 1" cubes

1 lb shrimp, deveined and without shells

24 little neck clams

3/4 lb bay scallops

1/2 C fresh basil, (chopped)

2 tbsp lime juice (freshly squeezed)

Extra basil and lime wedges for garnish

Instructions:

Heat olive oil in large, heavy pot over medium heat.

Sauté the chopped shallots for about five minutes or until they turn translucent.

Throw in the garlic and continue to cook for two minutes.

Stir in the jalapeno pepper and the red pepper continuously stirring until the peppers begin to soften.

Put in the red curry paste, coconut milk, honey, clam juice, salt and fish sauce in that order. Keep on stirring.

Add the potato cubes and cook another fifteen minutes or until potato cubes are soft.

Stir in the cod, scallops and shrimps. Cover and continue to cook for another five minutes.

Put in the clams next and again cover, and cook until all the clams have opened (about five minutes). Remove the pot from the fire.

Stir in the basil and lime juice.

Decorate with a lime wedge garnishing and a sprinkle of basil.

Hot and Sour Chicken Soup

Ingredients:

2 tbsp Coconut Oil

1 Cup Onion, sliced

1/2 Cup carrots, diced

1/2 Cup Celery, sliced

2 Skinless Chicken Breast- medium diced

6 Cups Chicken Stock

5 Mushrooms, sliced

1 tbsp fresh ginger minced 3 cloves garlic

3 tbsp Coconut Aminos

1 1/2 tsp Honey

3 tbsp White Wine Vinegar 1 1/2 tsp Sesame Oil

2 tbsp Tapioca Starch dissolved in 0.5 cup chicken broth.

6 Eggs, beaten

2 Green Onions Chopped

1/2 Cup Bamboo Shoots thinly Sliced

Salt

Pepper

Hot Chili Oil

Instructions:

Sauté onions in a heavy pot over medium/high heat for 3 minutes.

Add the carrots and sliced onions. Continue to sauté until carrots become tender.

Pour in the chicken stock and wait till it boils.

Throw in the mushrooms, ginger, coconut aminos, garlic, vinegar, honey, and sesame oil in that order.

Increase the temperature to bring the mixture to a boil.

Vigorously whisk the tapioca starch with the chicken broth and pour in the tapioca solution into the stock pot with the soup while stirring the soup continuously.

Allow the soup to boil again.

Continue to stir until the soup is thick then reduce the heat to medium.

Drizzle the whisked eggs slowly over the boiling soup while stirring continuously.

Add the green onions, bamboo shoots, pepper and salt together with a small amount of the chili oil. Reduce the heat to Low this time and let it simmer for 5 minutes.

Taste the soup. Add more chili oil if needed just a little at a time until the taste is spicy enough for you.

Holiday Bouillabaisse

1 can diced tomatoes, 28 oz (or you can use chopped fresh tomatoes instead)

1 onion, medium, chopped

½ large or 1 medium red bell pepper

3 to 4 stalks celery, trimmed and chopped

2 cloves garlic, chopped

½ lb shelled medium shrimps, deveined ,

¼ lb. swordfish, skinned and cut in cubes

½ lb. bay scallops

½ tsp cinnamon,

1 tsp cumin,

cayenne pepper

black pepper

salt,

2 cups water

Instructions:

You should have approximately equal volumes of chopped onions, bell peppers and celery. Add up in any if needed.

Sauté onion, garlic, bell pepper and celery, in 3 tablespoon olive oil.

Add black pepper and salt.

Add the swordfish cubes and continue to cook until the vegetables are tender but firm.

Add the tomatoes, two cups of water, cinnamon and cumin.

Put a dash or two of cayenne pepper according to your taste. Boil for five minutes.

Add the shrimp and keep on boiling for another 4 minutes. Stir in the scallops and boil for another minute this time.

Reduce to low heat. Cover the pan and allow to simmer for not less than 20 minutes and no more than 25 minutes.

Decorate with fresh parsley sprigs before serving.

CPSIA information can be obtained
at www.ICGtesting.com
Printed in the USA
LVOW13s0209260318
571148LV00014B/312/P